BIOSYSTEMATIC STUDIES OF CONIFER-FEEDING CHORISTONEURA

Biosystematic Studies of Conifer-feeding Choristoneura (Lepidoptera: Tortricidae) in the Western United States

Edited by Jerry A. Powell

UNIVERSITY OF CALIFORNIA PRESS
Berkeley • Los Angeles • London

UNIVERSITY OF CALIFORNIA PUBLICATIONS IN ENTOMOLOGY

Volume 115
Issue Date: December 1995

UNIVERSITY OF CALIFORNIA PRESS
BERKELEY AND LOS ANGELES, CALIFORNIA

UNIVERSITY OF CALIFORNIA PRESS, LTD.
LONDON, ENGLAND

PRINTED IN THE UNITED STATES OF AMERICA

Library of Congress Cataloging-in-Publication Data

Biosystematic studies of conifer-feeding Choristoneura
(Lepidoptera: Tortricidae) in the western United States / edited by
Jerry A. Powell.
p. cm. — (University of California publications in
entomology ; v. 115)
Includes bibliographical references.
ISBN 0-520-09796-3 (pbk : alk. paper)
1. Choristoneura—West (U.S.)—Classification. 2. Conifers—
Diseases and pests—West (U.S.) I. Powell, Jerry A. II. Series.
QL561.T8B56 1996
595.78'1—dc20 95-41081
CIP

The paper used in this publication meets the minimum requirements of American National Standard for Information Sciences—Permanence of Paper for Printed Library Materials, ANSI Z39.48-1984.

Contents

Acknowledgments, vii

1. Introduction: Historical Review, Taxonomic Problems, and Research Approaches
J. A. Powell 1

2. Biological Relationships: Host Tree Preferences and Isolation by Pheromones Among Allopatric and Sympatric Populations of Western *Choristoneura*
J. A. Powell and *J. A. De Benedictis* 21

3. Pheromone Chemistry and Response Behavior of Conifer-feeding *Choristoneura* in the Western United States
G. E. Daterman, L. L. Sower, R. E. Stevens, and *D. G. Fellin* 69

4. Phenetic Studies of Spruce Budworm Populations (*Choristoneura* Species)
J. A. De Benedictis 85

5. Studies of Pheromone Attraction and Inheritance Using Virgin Females of *Choristoneura* Species, in the Central Sierra Nevada, California
J. A. De Benedictis, A. M. Liebhold, and *J. A. Powell* 151

6. Foliage-feeding Lepidoptera of *Abies* and *Pseudotsuga* Associated with *Choristoneura* in California
J. A. Powell and *J. A. De Benedictis* 167

7. Evolutionary Interpretation, Taxonomy, and Nomenclature
J. A. Powell and *J. A. De Benedictis* 217

Acknowledgments

We thank the people who were instrumental in development of this study: Nils Westerland, Tahoe City, California, carried out blacklight trapping during 1974-1981, providing flight period and relative abundance data for *Choristoneura retiniana* and *C. lambertiana subretiniana*. He and Florence Westerland cooperated by monitoring pheromone traps and providing us with hospitality during the 1979-81 field seasons. R.C. Heald and other authorities in the former Department of Forestry, University of California, Berkeley, arranged for us to use facilities at the Blodgett Forest Research Station. W.J.A. Volney, formerly of the Department of Entomology, UC Berkeley, provided considerable insight into *Choristoneura* biology, techniques and forest entomology practices. Research by Volney's group in southern Oregon was instrumental in the development of our study. Cooperation by curators of several museum collections, acknowledged by De Benedictis (1995, Appendix 4A), enabled study of specimens in their care. Funding from the USDA, CANUSA-West Program supported our research during parts of 1978-79 and 1980-81. Grant No. BSR-8411459 from the National Science Foundation supported parts of this study during 1985-88. Finally, we sincerely thank five anonymous reviewers, who provided insightful criticism of various combinations of the parts of the series. While some of their comments were contradictory to one another, we believe that revisions based on the queries have greatly improved the text.

1

Introduction

Historical Review, Taxonomic Problems, and Research Approaches

Jerry A. Powell
Essig Museum of Entomology
University of California, Berkeley, California

Abstract

Members of the Spruce Budworm complex, conifer-feeding tortricids related to *Choristoneura fumiferana* (Clemens), are the most extensively studied moths in the Western Hemisphere, yet taxonomic relationships among western populations have remained poorly understood. Varying degrees of reproductive isolation exist; morphological differences are subtle and the characters vary discordantly, some expressing polymorphism, some varying clinally; and nomenclatural confusion arose because taxonomists tried to interpret species on the basis of too few specimens with scant knowledge of biological features.

In addition to *C. fumiferana,* which was broadly used for western populations prior to 1967, 12 species group names have been applied to budworms in the western United States. The earliest of these, *C. retiniana* (Walsingham), 1879, and *C. lambertiana* (Busck), 1915, were assigned to different genera; it was not until the 1930's that all members of the complex were recognized as members of a single genus, and not until 1947 that they were properly assigned to *Choristoneura.* Even as late as the 1960's, available information did not permit realistic interpretation of relationships. Taxonomists tried to use genitalia characters to define species in the belief that there must be distinguishing morphological features by which individual moths can be assigned to taxa; none exists. This introduction reviews the history of the changing nomenclature and development of taxonomic concepts.

The discovery of pheromone chemistry differences between species and subsequent identification of their principal components was the key to understanding relationships. We proposed three objectives for these studies: (1) to define biological relationships between sympatric pairs of the Abietoideae- and Pinoideae-feeding populations; (2) to document host tree preferences in larval feeding and geographical distributions of those associations; and (3) to conduct detailed morphological studies leading to a useful taxonomic frame work and nomenclature. We employed several approaches: trapping moths in natural populations by ultraviolet lights and testing attraction of males to synthetic pheromone baits and to virgin females obtained from larval rearing; field collection and subsequent rearing of larvae; and laboratory study of morphological characters in population samples defined by pheromone attraction and hostplant preferences.

INTRODUCTION

The Spruce Budworm, *Choristoneura fumiferana* (Clemens), is the most important defoliator of coniferous forest trees in North America, particularly in Canada. Probably no species of Lepidoptera has been studied more intensively. I noted 30 years ago that more than 50 papers had been published on this insect during the preceding 15 years (Powell 1964:171)—that boast now seems absurd. McKnight *et al.* (1988) cite more than 2,500 references to the Spruce Budworm, among 4,300 titles published through 1985 that mention some aspect of the species complex. Probably we know more about the hostplant selection, growth, overwintering survival rates, pheromone chemistry, consequences of control measures, sampling methods of all stages, and parasites and predators of this moth than of any other in North America, if not the world. Yet in parts of its range, we do not know its name.

There are no clearly definable characters to distinguish named entities that comprise the complex of so-called species. For example, the Western Spruce Budworm, *C. occidentalis* Freeman, a species that was not described until 1967, is not well defined because the name has been applied to differing populations over widespread parts of the western Nearctic—yet more than 450 publications refer to this name.

The reasons for this seeming paradox arise from three interrelated sources. First, and most important, are the underlying biological relationships among populations: varying degrees of reproductive isolation exist. Members of two species will mate and produce viable offspring in laboratory confinement (e.g., Smith 1953, Campbell 1967), although reciprocal crosses among most named entities in the western United States have not been tested.

Second, morphological differences are subtle and vary discordantly within and between populations and do not consistently reflect species separation as defined by pheromone chemistry. Many populations exhibit extreme ranges of variation, either continuous or polymorphic or both, in size, color, or morphological details in adults, larvae, and pupae. Other populations are relatively monomorphic. Allopatric polytypy and clinal variation exist, and hostplant preferences vary geographically, among populations that share the same pheromone chemistry. The characters conventionally used by moth taxonomists in defining species, structures of the genitalia, are confoundingly similar among all members of the complex; and some genital features vary independently within and between populations.

Finally, confusion arose because taxonomists applied names prematurely, trying to interpret species relationships on the basis of too few specimens, with little or no knowledge of biological features. In some instances taxonomists acted without carefully considering the existing names and their application. They doggedly tried to force genitalia characters to define species in the belief that there must be distinguishing morphological features by which individual moths can be assigned to taxa. None exists.

In this discussion and in the treatments that follow, we use the following interpretations of terms:

population: all the individuals in a given geographic area that could interbreed (i. e., dispersal capabilities and sex pheromone production/response potentially allow indivdual males and females to meet, mate, and produce viable progeny).

form, morph, polymorph form: an individual or set of individuals within a population that show distinctive phenotypic characteristics.

race: any set of populations that is geographically limited within a species, defined by morphological and/or biological similarity.

subspecies: any population or set of populations within a species that has been given a formal name meeting the criteria of the International Code of Zoological Nomenclature. The nominate subspecies encompasses all populations not otherwise restricted to one of the other subspecies.

species: the total set of populations that we believe consists of potentially interbreeding individuals. In our studies, bioassay of sex pheromone response is the criterion used to define "potentially interbreeding."

HISTORICAL REVIEW

The Descriptive Era

Most North American microlepidoptera were described between 1860 and 1920, when species concepts were typological and discrimination was based on external features, prior to the time when structures of the genitalia became the basis for generic and specific classifications. About 70% of named, native Tortricinae in the United States and Canada were described prior to 1920. Closely related species often were placed in different genera, varying with concepts of a given era, and their relationships went unrecognized for many years. Such was the nomenclatural fate of the spruce budworms.

The earliest name applied to a North American conifer-feeding *Choristoneura* is *fumiferana* (Clemens 1865), which was described in the Linnaean genus *Tortrix*, from "Virginia." Next, a form later recognized as the dark color phase of *fumiferana* was described as *Tortrix nigridia* (Robinson 1869), from "Ohio, Pa., and Mass." Its synonymy was recognized early (Fernald 1881), and no further confusion with its identity or additional synonyms of *fumiferana* have been added.

The next oldest name for a species later considered to be a conifer-feeding *Choristoneura* is *houstonana* (Grote 1873), which also was described in *Tortrix* and was based on a specimen from Texas. Another name was applied to the same species, *Lozotaenia retana* (Walsingham 1879), also from Texas. This too was soon synonymized (Grote 1881), and no further nomenclatorial confusion ensued. The species was transferred to *Choristoneura* by Freeman (1958), then assigned to a new genus, *Cudonigera* (Powell and Obraztsov 1977), expressing its marked morphological dissimilarity from species of *Choristoneura*. The taxonomy, biology, and geographical distribution have been discussed elsewhere (Powell 1964, Heinrichs 1971, Heinrichs & Thompson 1968, Powell and Obraztsov 1977). The larvae feed on Cupressaceae, while all members of the *C. fumiferana* complex depend upon Pinaceae. Hence, it is not considered further in this study.

The earliest name applied to a western budworm of the *C. fumiferana* complex is *C. retiniana* (Walsingham 1879), which was assigned to *Lozotaenia*. This species was described from one specimen from Mt. Shasta in northern California. The type specimen was collected in August 1871, when the Walsingham expedition stayed near Mt. Shasta for a month. Forays by horseback were made from the base camp (Carrier diary at BMNH and copy at UCB, Essig 1941), so the precise locality and elevation of the *C. retiniana* collection are unknown. We now know that there are sympatric *Pseudotsuga*- and *Abies*-feeding species on the lower slopes of Mt. Shasta. Moreover, the fir-feeding population there, to which the name *C. retiniana* is ascribed, shows intermediate characteristics between the green budworms of the Sierra Nevada and those to the north.

Later *Lozotaenia retiniana* was placed in *Archips* (=*Cacoecia*) (Dyar 1903). Meanwhile, however, *fumiferana* and *houstonana* remained in *Tortrix* with a diverse conglomeration of species, members of three tribes according to present concepts. Meyrick (1912) transferred *fumiferana* to *Harmologa*, a New Zealand genus and left *houstonana* in *Tortrix*. This arrangement was followed by Barnes and McDunnough (1917) and Forbes (1923), who influenced the American literature for the following two decades.

The oldest name applied to a pine-feeding population in Nearctic *Choristoneura* is *Tortrix lambertiana* Busck (1915), described from Ashland in southern Oregon. The species had been reared from Sugar Pine at the Ashland Forest Service Station (Hopkins data given by Powell 1964:178), and the precise area of collection is unknown.

The final name proposed during the early descriptive era was *Tortrix carnana* Barnes and Busck (1920), which was applied to the distinctive rust and white colored moth that we now know to be associated with Bigcone Spruce in southern California. The type locality, which was given as "Camp Baldy, San Bernardino Mts.," is believed to be near Mt. Baldy P.O. in the San Gabriel Mountains (Powell 1964, 1980).

Thus, by the 1920's, names had been given to the major components of the Nearctic conifer-feeding budworm complex, without comprehension of their relationships. Only two, the Spruce Budworm, "*Harmologa*" *fumiferana*, and the Sugar Pine Tortrix, "*Tortrix*" *lambertiana*, had been reared from their coniferous hosts. None had been recognized as related to the Palearctic *Choristoneura diversana* (Hübner), the type species of *Choristoneura*. The genus was described by Lederer in 1859, but it was considered to be a synonym of *Tortrix* in the early 1900's (Meyrick 1912).

Era of Synthesis, 1921-1953

Pierce and Metcalfe (1922) made the major advance to bring natural order to tortricid moth classification, when they illustrated the genitalic structures of the British species. They placed species in nine groups that essentially resemble tribes of modern classifications. The type species of *Tortrix*, *viridana* L., was recognized as markedly distinct from those of the archipine genera; and *diversana*, the type species of *Choristoneura*, was transferred from *Tortrix* to *Archips* (=*Cacoecia*). While this work did not have an immediate impact on American nomenclature, McDunnough (1939) reassigned many species by the Pierce and Metcalfe system, and for the first time brought together all the described budworms in one genus, *Archips*.

During this period, concern grew about massive outbreaks of budworms, especially in eastern Canada during 1910-1922 (Swaine 1922, Swaine and Craighead 1924), in Minnesota on pines during the 1920's (Graham 1925, 1929), and again in Ontario starting in 1937 (Elliott 1960). Intensive studies of population dynamics and other biological aspects ensued and ultimately led to more critical taxonomic investigations at both the specific and generic levels.

An important product of this period was the gradual recognition of a sibling species associated with pine in eastern North America. The pine-feeding budworm was recognized as an ecological variety of *fumiferana* by Graham (1929). More detailed analyses by Brown and MacKay (1943) demonstrated subtle differences in size, wing color, genital morphology and phenology. Research on the separation and definition of the pine-feeding species was culminated by Freeman (1953), MacKay (1953), Campbell (1953), Cox (1953), and Smith (1953), when *Choristoneura pinus* Freeman was formally named.

Freeman (1947) applied the name *Choristoneura* to the Nearctic budworm complex and transferred *C. fumiferana* to that genus. Although he erroneously cited *C. rusticana* (Treitschke) as the type species, he must have compared the genitalia of *C. diversana*, the true type species, in order to have perceived the relationship of *C. fumiferana*. The Palearctic *rusticana*, which is the type species of *Clepsis* Guenée, is structurally quite different. Later, Freeman (1958) cited the two type species correctly and interpreted the American members of their respective genera accurately.

Thus by 1953, taxonomists and forest biologists recognized two species in the budworm complex, *Choristoneura fumiferana* and *C. pinus*, which are broadly sympatric in eastern North America, and at that time all of the Abietoideae-feeding populations of western regions were included within the concept of *C. fumiferana*. Subsequently, concepts of relationships between *C. fumiferana* and *C. pinus* in the eastern areas have remained stable, while several older species names that had not been associated with *Choristoneura* and several newly proposed species and subspecies names have been applied to western populations in this complex.

Classical Taxonomic Treatments, 1954-1964

Obraztsov (1954, 1955) redescribed *Choristoneura*, cataloged the Palearctic species, and confirmed the generic placement of *C. fumiferana*. Freeman (1958) attempted to define and classify all the North American species of *Choristoneura*. He transferred *retiniana, carnana*, and *lambertiana* to this genus, recognizing them as closely related to *C. fumiferana*. He presented a key to species based on forewing color pattern, redescribed the adults, and illustrated them by black and white photographs. His stylized drawings of male genitalia show apparent differences between the species, but he reported neither the geographical data for specimens illustrated nor number of dissections examined. The differences portrayed in form of the valva, transtilla, anellus, and aedeagus reflect individual rather than species distinctions and evidently are the result of preparation technique combined with too few dissections examined, and/or poor artistic interpretation. Females were not distinguished.

The specimen that Freeman illustrated as *C. retiniana* is a male of the pine-feeding complex, later named *C. subretiniana* Obraztsov (1962), from Mono Co., California. Illustrated specimens cited as "*C. fumiferana* complex" include a male from New Lisbon, N.J., later described as *C. pinus maritima* Freeman; males from Colorado later ascertained to be pine-feeders, *C. lambertiana ponderosana* Obraztsov; and the tan form of *C. retiniana* from Modoc Co., California. There is no evidence that Freeman saw the type specimen of *C. retiniana* or other material that he recognized as this species from northern California, except for the Modoc green budworm population, which he assigned to "*C. fumiferana* complex." Instead, his characterization of *C. retiniana* appears to have been based on the pine-feeding *C. subretiniana*. Hence, neither the identities of the specimens with which Freeman worked nor their distinguishing characteristics were correctly determined.

Freeman reported *C. lambertiana* from California and New Mexico; the latter may have been based on a pine-feeding population, but his concept of *C. lambertiana* did not include the other Rocky Mountains pine associates. He also reported *Juniperus* as a hostplant of *C. lambertiana*. Possibly this record originated from a misidentified specimen of *Cudonigera houstonana* (Grote) or the superficially similar *Argyrotaenia cockerellana* (Kearfott). The latter was omitted from Freeman's revision of the tribe, so it was not a species that he recognized. Both are restricted to *Juniperus* so far as known (Heinrichs 1971, Powell 1964 and unpubl. data, Powell and Obraztsov 1977).

Obraztsov began a study of the Nearctic genera of Tortricinae, and during 1956-1966 he made considerable progress in clarifying the poorly known North American fauna. He also undertook species treatments in a number of Nearctic and Neotropical genera. He was naive about the biogeographic complexity of the western states, and like Freeman did not use available specimens from western collections, so he did not appreciate the extent of variability within and between populations.

In *Choristoneura*, Obraztsov (1962) attempted to clarify the identity of *C. retiniana* and differentiate it from several other western members of the complex, including four new taxa. The treatment was not comprehensive in that he used only the American Museum of

Natural History (AMNH) collection and selected specimens from the U.S. National Museum of Natural History (NMNH), and he excluded diagnosis of *C. fumiferana*, *C. carnana*, and *C. pinus*. He did not differentiate any western populations later called *C. occidentalis*, which were then referred to *C. fumiferana*. Obraztsov attempted to distinguish the remaining taxa on the basis of individual differences in uncus shape, socii length, and aedeagus form, and in the females he illustrated perceived differences between *C. lambertiana* and what he understood to be *C. retiniana*, in the shape of the sterigma and antrum.

Obraztsov studied the type specimen of *C. retiniana* at the British Museum of Natural Histroy (BMNH) and illustrated the genitalia, confirming Freeman's (1958) generic placement. Unfortunately, the genitalia had been dissected many years previously and retained as a dry mount. Obraztsov rehabilitated the preparation by recleaning, staining, and mounting on a slide, but the results are less than satisfactory, with the structures flattened, distorted, and overstained. Even if individual morphological differences of the sort Obraztsov illustrated could be discerned between species in this complex, they could not be seen on this preparation. Hence, Obraztsov's diagnosis is based on specimens from other localities. Although he believed that specimens he compared with Walsingham's type were conspecific, his concept of *C. retiniana* was based on an imponderable conglomeration of six individuals from four widely scattered, unrelated populations: a paratype male of *C. carnana californica* Powell, description of which was in press concurrently; a female of the Sierra Nevada population of *C. retiniana* from Yosemite Valley; and samples of two populations that presumably are *C. occidentalis* according to present concepts: two males and a female from northeastern Oregon, and two females from southern New Mexico. Most of his diagnostic illustrations are based on the specimens from Oregon. Thus the basis for differentiation of new taxa from the oldest named western species was erroneous.

Obraztsov proposed four new names, two as species and two as subspecies: *C. spaldingiana* was based on a single, worn male from Provo, Utah, collected in 1907; *C. subretiniana,* from high elevation, east slope Sierra Nevada; *C. lambertiana ponderosana* from the eastern flank of the Rocky Mountains in Colorado; and *C. lambertiana lindseyana* from Modoc Co., California.

Pheromone trapping in 1981 demonstrated that *C. spaldingiana* is the mottled form of the dimorphic *C. retiniana* populations that occur in the Wasatch Mountains, Great Basin Ranges, Sierra Nevada, and mountains of southern California (Powell and De Benedictis, Part 2 this vol.). This relationship was suggested by Obraztsov (1962), who cited two specimens from Fallen Leaf Lake near Lake Tahoe as possibly conspecific with *C. spaldingiana*. However, he failed to perceive the relationship of the tan form in the Sierra Nevada population, specimens of which he referred to *C. lindseyana*.

Choristoneura subretiniana was described as a species separate from *C. lambertiana*. The type series had no hostplant association, but there are specimens from Modoc Co., California, in the NMNH that are phenotypically similar to typical *C. subretiniana*, some of which were reared from Lodgepole Pine in 1952-53 (Powell, 1964:183); these were ignored by Obraztsov.

Conversely, *C. lambertiana ponderosana*, which is phenotypically dissimilar from typical *C. lambertiana*, was proposed by Obraztsov as a subspecies, evidently primarily on the basis of the pine association. He examined material representing the more variable, pine-feeding populations of Montana, Idaho, and Wyoming and assigned them to *C. lambertiana lambertiana*; but he did not attempt to assess relationships of *C. l. ponderosana* to them. In addition, he assigned one of two females taken together near Cloudcroft, N.M., to *C. lambertiana* s. str. (the other to *C. retiniana*).

Finally, Obraztsov imagined the tan morph of *C. retiniana* to be a subspecies, which he named *C. lambertiana lindseyana*. The type series was selected from a large collection (ca. 200 specimens) of *C. l. subretiniana* and the variably patterned morphs of *C. retiniana* at the NHNM that had been collected in the Warner Mountains, Modoc Co., by A.W. Lindsey in 1922. The reticulate forms of that population were not included in Obraztsov's treatment. A collection from Fallen Leaf Lake near Lake Tahoe included two specimens listed under *C. lindseyana*, while mottled ones were mentioned under *C. spaldingiana*.

My assessment (Powell 1964), which was in press concurrently with that of Obraztsov (1962), was limited to the species represented in California but involved study of all western Nearctic populations. Although I examined material in most of the western U.S. collections as well as the major eastern museums (AMNH; Canadian National Collection, CNC; NMNH), a fragmentary picture was presented by available specimens. Exclusive of the type series of *C. carnana*, *C. c. californica*, and *C. l. subretiniana*, and the Warner Mountains populations, there were fewer than 90 specimens representing the entire Pinaceae-feeding *Choristoneura* complex in California. These originated from 40 localities, 28 of which were represented by single specimens. Moreover, there were no host tree associations documented by adults reared from larval collections except from the pine- and White Fir-feeding populations of the Warner Mountains.

The first step in unraveling the nomenclatural tangle surrounding the western budworm species was to correctly interpret the oldest name, *C. retiniana*, and associate its identity with extant populations. I did not see Walsingham's type specimen, but on the basis of Obraztsov's photographs of it I surmised that it represents a population of fir-feeding green budworm similar to those of Modoc County. The similarity of specimens from the Warner Mountains to a photograph of the type of *C. retiniana* convinced me that the Modoc population is conspecific. There was too little evidence to characterize *C. retiniana* as distinct from *C. fumiferana*, which at the time included all populations subsequently referred to *C. occidentalis*, and therefore I treated *C. retiniana* as a synonym. Through specimens reared by U.S. Forest Service personnel in 1950-53 and by me in 1957, I perceived that the tan form of *C. retiniana* represents a polymorph form and relegated its name, *C. lindseyana*, to synonymy.

I associated populations of the west slope of the Sierra Nevada and the North Coast Range (*californica*) with the typical race of *C. carnana* in southern California, but there were no larval records to show that *C. carnana* is a Douglas-fir-feeder. I saw the curious circumstance that these populations are allopatric to the known pine-feeders, *C. l. lambertiana* in the Siskiyou Mountains and *C. l. subretiniana* in the Warner Mountains and high east side Sierra Nevada. This led to the erroneous supposition that the whole ring of populations comprises a single pine-feeding complex or superspecies. Hence, I assigned *carnana* as a subspecies of *C. lambertiana* and described the North Coast Range component as *C. lambertiana californica.* Had there been a collection demonstrating sympatry of *C. carnana* and *C. lambertiana* (including *subretinana*), my notion of a single pine-feeding polytypic species would have been untenable.

I suspected that the unmarked tan forewing phenotype from scattered Sierran localities was a form comparable to that of the Warner Mountains, but there was insufficient material to convince me that it was conspecific with *C. retiniana.* Two specimens of the tan form from the San Bernardino Mountains in southern California led me to suspect that the assignment of *C. carnana* to the same species as the dimorphic populations was unlikely (Powell 1964:183), but there were no collections of sympatric *C. retiniana* and *C. carnana* at the time.

As a result, in the early 1960's concepts of species in western conifer-feeding *Choristoneura* and the characters employed to distinguish them were artificial. Our *C. "fumiferana"* included the complex populations later to be designated *C. occidentalis* and *C. retiniana*, while *C. "lambertiana"* included *C. carnana*, some Sierra Nevada *C. retiniana* and Coast Range *C. californica.*

Integration of Biological Characteristics

Several discoveries during 1964-1974 set the stage for the present series of studies. These were achieved independently by the Canadian Department of Forestry, USDA Forest Service, and our group at Berkeley. The most important of these was the discovery of pheromone chemistry-based differences between species (Sanders 1969, 1971a, b, c) and the subsequent identification of principal components of the pheromones, enabling production of synthetic attractants (Sanders *et al.* 1974, Weatherston *et al.* 1971).

G.W. Stehr and his associates at the Forest Research lab, Sault Ste. Marie, Ontario, carried out an intensive study of the biology, including genetic and physiological aspects, of Canadian and western U.S. populations of budworms. This project began with collections in the 1950's, but the comprehensive survey of western populations took place in 1961 and 1962 (Harvey & Stehr 1967). The material consisted of 287 collections of late instar larvae, about half of which were from western areas. Most were taken from populations at epidemic densities. The larvae were shipped to Sault Ste. Marie and reared under controlled conditions.

Diverse studies were carried out with the progeny of the resultant lab colonies (Freeman 1967, Freeman and Stehr 1967, Stehr 1967, Harvey and Stehr 1967, and Campbell 1967. Freeman described and provided names for four new species: *C. occidentalis,* applied to Pacific Northwest and Rocky Mountain populations; *C. orae*, a small, dark form in north coastal British Columbia; *C. biennis* for two-year cycle populations in higher elevation parts of interior British Columbia; and *C. viridis,* the Warner Mountains *Abies*-feeding species. These names replaced use of *C. fumiferana* for western populations. There were two unfortunate decisions in application of these names: (1) the type locality of *C. occidentalis*, near Yakima, Washington, is at the northern edge of a zone of hybridization, and we do not yet know the extent to which variability in that population is affected, and (2) *C. viridis* was applied to the same population as *C. lindseyana* Obraztsov, and therefore is an obvious synonym. Freeman ignored the subjective synonymy of *C. lindseyana* that I had proposed (Powell 1964). He evaded concern about nomenclatural priority by presuming that all previously named western budworms (i.e., *C. retiniana, C. lindseyana, C. spaldingiana, C. carnana, C. californica, C. lambertiana, C. subretiniana,* and *C. ponderosana*) feed on *Pinus*, even though the first two had been recorded from *Abies* (Powell 1964). These eight named *Choristoneura* were excluded from his treatment (Freeman 1967, Freeman and Stehr 1967). The pine-feeding budworms, *C. lambertiana* and *C. l. subretiniana*, were interpreted as two species by Stehr (1967) and Harvey and Stehr (1967), who ignored the names *C. carnana, C. lindseyana, C. spaldingiana,* and *C. l. ponderosana*, although clearly they were dealing with at least the last.

The Canadian studies improved our understanding of biological relationships among western budworms considerably. Application of the names, however, further confused matters, especially because the oldest name, *C. retiniana*, which was known to represent *Abies*-feeding populations (Powell 1964), was disregarded, and because *C. occidentalis* and *C. lambertiana* were applied to widespread populations that exhibit considerable variation within each species.

In California, several discoveries during the 1960's and early 1970's further clarified the picture. These data helped identify *C. retiniana*, showed that there are two sympatric Abietoideae-feeding species on the west slope of the Sierra Nevada, and provided records of hostplant associations for *C. retiniana*, *C. carnana*, and *C. subretiniana.*

In 1962 C.D. MacNeill and D.C. Rentz collected *Choristoneura* at McBride Springs on the west side of Mt. Shasta, at 1,400 m in the *Pseudotsuga-Abies concolor* zone. They obtained specimens that resemble the type of *C. retiniana* from Mt. Shasta. P.A. Opler and I made another collection at McBride Springs, and I compared specimens from this population at the BMNH in 1971 and found that some individuals are essentially indistinguishable from Walsingham's type.

In 1966 I surveyed at Mt. Ashland, ca. 90 km northwest of Mt. Shasta near the Oregon border, and collected a series of *Choristoneura* that includes a diversity of forms not readily assignable to any species. The following year, I sampled moths at Blodgett Forest, El Dorado Co., at 1,300 m on the west slope of the Sierra Nevada, during a five-week period, and found two obviously different *Choristoneura* in sympatry, one *C. carnana*-like and the other showing dimorphism of the Sierran forms Obraztsov had associated with the names *C. lindseyana* (tan) and *C. spaldingiana* (mottled).

R.E. Dietz, P.A. Rude, and I attempted a survey for larvae during June and July 1970 at Mt. Ashland, Mt. Shasta, and southward in the Sierra Nevada. Although results were negative in the northwestern areas, we did rear *C. retiniana* from *Abies concolor* at five sites in the Warner Mountains and Sierra Nevada, including specimens resembling *C. spaldingiana* in the vicinity of Truckee, California. Then in 1974, T.D. Eichlin and I reared typical *C. retiniana* from *A. concolor* at 1,850 m on Mt. Shasta, a higher elevation than *Pseudotsuga* occurs there. This confirmed the host tree association of *C. retiniana* s. str.

Referral of *C. subretiniana* to a race of *C. lambertiana* was inferred by rearing records of a phenotypically similar population in the Warner Mountains, Modoc Co. (Powell 1964). This association was confirmed in 1963 by discovery of larvae of *C. l. subretiniana* feeding on Lodgepole Pine in the Sierra Nevada (Stark and Borden 1965). A critical step toward clarification occurred in 1970, when there was a minor outbreak of *Choristoneura* on *Pseudotsuga menziesii* near Angwin, Napa Co., and T. Koerber of the USFS reared specimens of *C. californica*. This suggested that populations of typical and Sierran *C. carnana* were also taxa that feed on Douglas-fir, rather than pine-feeders as previously supposed.

During a massive outbreak of *C. retiniana* in the Warner and Manzanita mountains, Modoc Co., in 1974, the USFS attempted to monitor the effects of dylox spray for control. They subcontracted analysis of pre- and post-spray samples of various arthropods from nine localities Subsamples of 100 larvae of Green Budworm from each site and all the larvae judged to be other Lepidoptera in initial screening were sent to Berkeley, from each of three sampling periods. We reared more than 5,300 moths, 60% of which were *Choristoneura* (Powell and Donahue 1974). This better documented variation in *C. retiniana*, confirmed constancy of host tree preference of *C. l. subretiniana*, and elucidated the associated caterpillar community.

TAXONOMIC PROBLEMS AND OBJECTIVES

I resumed analysis of the taxonomy in 1978, in collaboration with G.E. Daterman, who planned to investigate the pheromone chemistry of western budworm populations, and R.E. Stevens, who would carry out biological studies in the southern Rocky Mountains.

Taxonomic problems consist of our unsatisfactory attempts to apply a rigid system of categories and names to sets of populations, which, in *Choristoneura*, may not exist in nature as genetically discrete species and objectively definable subspecies. It had become increasingly obvious that morphological study of additional specimens was not going to resolve our understanding of the species complex. A few strictly nomenclatural problems and questions relating to too few available specimens persisted, but the primary difficulty was our lack of understanding of the insects' biological interrelationships, their degrees of reproductive isolation, and the constancy of their host tree preferences. Resolving these questions was necessary to progress further.

We proposed the following objectives:

1. To define biological relationships between sympatric pairs of Abietoideae-feeding *Choristoneura* and between these and the Pinoideae-feeding species of *Choristoneura*.

2. To further document geographical distributions and host tree associations, particularly for previously unknown or poorly known populations and in areas where questions of taxonomic identity exist.

3. To conduct morphological studies to define structures useful in identification of taxa, and establish a taxonomic framework based on biological relationships, which will enable predictive qualities for less well known populations, and provide a valid nomenclature for communication purposes.

Following are the primary problems that we addressed during 1978-1982, in seeking to establish criteria for defining species, by determining reproductive isolating mechanisms (pheromone chemistry and behavioral responses).

Choristoneura retiniana: Clarification of the relationships of the Mt. Shasta population to those of *C. carnana* in the vicinity and to *C. occidentalis* to the north. Scattered specimens from Siskiyou Co. and those from Mt. Ashland could not be assigned to any taxon with confidence. Moreover, the relationships of the Siskiyou-Modoc *C. retiniana* to populations of the Sierra Nevada and ranges of southern California were not clear. Most southern populations were represented by only one or a few specimens that showed differences but also similarities to those of Modoc Co.

Choristoneura carnana: Confirmation of the *Pseudotsuga*-feeding habit discovered in Napa Co. in 1970 and its constancy in the Sierra Nevada and southern California. This would refute the association with pine-feeding *C. lambertiana*, and would provide supporting biological grounds for the sympatry of two Abietoideae-feeding forms in the Sierra Nevada. Determination of reproductive isolation of these two entities by pheromones was a key to our understanding Californian components of the complex.

Choristoneura occidentalis: Western populations assigned to this name that respond to the E-11 tetradecenal synthetic pheromone show geographic differences in size and color of adults and larvae and in hostplant preferences. The interpretation by Freeman (1967) and Stehr (1967) implies that *C. occidentalis* is sympatric, or nearly so, with *C. retiniana* (=*viridis*) in northeastern California. Hence, study of populations in northern California and Oregon was a primary goal.

Choristoneura lambertiana: There were questions as to which groups of populations and subspecies should be included within a species concept under this name. There is pronounced geographical variation in adult phenotype and the correlated host tree associations, between Siskiyou Mountains *C. l. lambertiana* (on *Pinus lambertiana*), Sierra Nevada and Warner Mountains *C. l. subretiniana* (on *P. contorta murrayana*), and *C. l. ponderosana* of the Rocky Mountains (on *P. ponderosa*). Populations of Idaho, Montana, and Wyoming (on *P. contorta murrayana*, *P. flexilis*, and *P. ponderosa*) exhibit a range of forewing pattern showing elements of all three of the named subspecies, with those of eastern Oregon (on *P. ponderosa*) phenotypically somewhat intermediate between Montana forms and *C. l. subretiniana*.

In general, however, we regarded the Abietoideae-feeding complexes to be the primary emphasis of the study, and the *C. lambertiana* complex was investigated secondarily, where sympatric with *C. carnana* or *C. occidentalis*.

RESEARCH APPROACHES

1. Survey by blacklight attraction and synthetic pheromone trapping to determine geographical distributions, seasonal flight periods, and potential host tree associations.
2. Search for larvae in natural populations and rear adults to document differential host tree preferences and to obtain virgin females to be used in field tests.
3. Trap males in natural populations in areas of known sympatric occurrence of two or more *Choristoneura*, using caged virgin females and/or synthetic pheromone lures.
4. Obtain living larvae, pupae or egg masses from larval collections or from confined females taken at blacklight, to establish laboratory colonies of selected populations for analysis of pheromone chemistry and protein esterases by collaborating investigators.
5. Conduct phenetic analyses of morphological characteristics that differ among populations defined primarily by pheromone attraction.

Results of the investigations outlined above are presented in the following papers and are summarized in Part 7.

LITERATURE CITED

Barnes, W., and A. Busck. 1920. Notes and new species. Contr. Nat. Hist. Lepid. N.A., 4:211-278, pls. 26-40.

Barnes, W., and J. McDunnough. 1917. Check List of the Lepidoptera of Boreal America. Herald Press, Decatur, Ill. viii + 392 p.

Brown, A.W.A., and M.R. Mackay. 1943. The jack pine budworm and the spruce budworm, *Cacoecia fumiferana* Clem. Canad. Entomol. 75:207-211.

Busck, A. 1915. Descriptions of new North American microlepidoptera. Proc. Entomol. Soc. Wash. 17:79-94.

Campbell, I.M. 1953. Morphological differences between the pupae and eggs clusters of *Choristoneura fumiferana* (Clem.) and *C. pinus* Freeman. Canad. Entomol. 85:134-136.

Campbell, I.M. 1967. On the coniferophagous species of *Choristoneura* (Lepidoptera: Tortricidae) in North America. IV. Sexual isolation between three species. Canad. Entomol. 99:482-486.

Clemens, B. 1865. North American micro-lepidoptera. Proc. Entomol. Soc. Phil. 5:133-147.

Cox, C.E. 1953. Analysis of frequency distribution of adults and larvae of *Choristoneura fumiferana* (Clem.) and *C. pinus* Free. Canad. Entomol. 85:136-141.

Dyar, H.G. 1903. A list of North American Lepidoptera and key to the literature of this order of insects. Bull. U.S. Natl. Mus. 52; 349 pp.

Elliott, K.R. 1960. A history of recent infestations of the spruce budworm in northwestern Ontario, and an estimate of resultant timber losses. For. Chron. 36:61-82.

Essig, E.O. 1941. Itinerary of Lord Walsingham in California and Oregon, 1871-1872. Pan-Pacific Entomol. 17:97-113.

Fernald, C.H. 1881. Oviposition in the Tortricidae. Am. Nat. 15:63-66.

Forbes, W.T.M. 1923. The Lepidoptera of New York and neighboring states. Primitive forms, Microlepidoptera, Pyraloids, Bombyces. Cornell Univ. Agric. Exp. Sta. Mem. 68; 729 pp.

Freeman, T.N. 1947. A new generic assignment for *Archips fumiferana* (Clemens), the spruce budworm. Canad. Entomol. 79:21.

Freeman, T.N. 1953. The spruce budworm, *Choristoneura fumiferana* (Clem.) and an allied new species on pine. Canad. Entomol. 85:121-128.

Freeman, T.N. 1958. The Archipinae of North America (Lepidoptera: Tortricidae). Canad. Entomol. 90 (Suppl. 7); 89 pp.

Freeman, T.N. 1967. On coniferophagous species of *Choristoneura* (Lepidoptera: Tortricidae) in North America. I. Some new forms of *Choristoneura* allied to *C. fumiferana*. Canad. Entomol. 99:449-455.

Freeman, T.N., and G. Stehr. 1967. On coniferophagous species of *Choristoneura.* (Lepidoptera: Tortricidae) in North America. VI. A summary of the preceding five papers. Canad. Entomol. 99:504-506.

Graham, S.A. 1925. Two dangerous defoliators of jack pine. J. Econ. Entomol. 18:337-343.

Graham, S.A. 1929. Principles of Forest Entomology. McGraw-Hill, New York. xiv + 339 p.

Grote, A.R. 1873. I. Description of new North American moths. Bull. Buff. Soc. Nat. Sci. 1:1-16.

Grote, A.R. 1881. "North American Tortricidae" by Lord Walsingham, London, 1879. Papilio 1:8-9.

Harvey, G.T., and G. Stehr. 1967. On coniferophagous species of *Choristoneura* (Lepidoptera: Tortricidae) in North America. III. Some characters of immature forms helpful in the identification of species. Canad. Entomol. 99:464-481.

Heinrichs, E.A. 1971. External morphology of larvae of *Choristoneura houstonana* (Lepidoptera: Tortricidae). Canad. Entomol. 103:12-17.

Heinrichs, E.A., and H.E. Thompson. 1968. The biology of *Choristoneura houstonana* (Lepidoptera: Tortricidae), a pest of *Juniperus* species. Canad. Entomol. 100:750-763.

MacKay, M.R. 1953. The larvae of *Choristoneura fumiferana* (Clem.) and *C. pinus* Freeman. Canad. Entomol. 85:128-134.

McDunnough, J. 1939. Checklist of the Lepidoptera of Canada and the United States of America. Part 2. Mem. So. Calif. Acad. Sci. 2(1):1-171.

McKnight, M.E., D.T. Jennings, S.C. Hacker, and F.B. Knight. 1988. North American coniferous *Choristoneura*: A bibliography. USDA For. Serv. Bibliogr. and Lit. Agric. 59; 580 pp.

Meyrick, E. 1912. Tortricidae. *In*: Wagner, ed. Lepidopterorum Catalogus. Pars 10; 86 pp.

Obraztsov, N.S. 1954. Die gattungen der palearktischen Tortricidae. I. Allgemeine aufteilung der familie und die unterfamilien Tortricinae und Sparganothinae. Tijdschr. Entomol. 97:141-231.

Obraztsov, N.S. 1955. Die gattungen der palaearktischen Tortricidae. I. Allgemeine aufteilung der familie und die unterfamilien Tortricinae und Sparganothinae. 1. Fortsetzung. Tijdschr. Entomol. 98:147-228.

Obraztsov, N.S. 1962. New species and subspecies of North American Archipini, with notes on other species (Lepidoptera: Tortricidae). Am. Mus. Novit. 2101:1-26.

Pierce, F.N. and J.W. Metcalfe. 1922. The genitalia of the group Tortricidae of the Lepidoptera of the British Islands.: Pierce and Metcalfe, Liverpool, England. xxii + 102 pp.

Powell, J.A. 1964. Biological and taxonomic studies on tortricine moths, with reference to the species in California (Lepidoptera: Tortricidae). Univ. Calif. Publ. Entomol. 32:1-317.

Powell, J.A. 1980. Nomenclature of Nearctic conifer-feeding *Choristoneura* (Lepidoptera: Tortricidae): Historical review and present status. Gen. Tech. Rep. PNW-100, USDA For. Serv.; 18 pp.

Powell, J.A., and J.A. De Benedictis. 1995. Biological relationships: host tree preferences and isolation by pheromones among allopatric and sympatric populations of western *Choristoneura*. Univ. Calif. Publ. Entomol. Part 2, this volume.

Powell, J.A., and V. Donahue 1974. Monitoring the effects of Dylox on microlepidopterous larvae. Rept. to USDA For. Serv., San Francisco; 17 pp. [unpublished].

Powell, J.A., and N.S. Obraztsov. 1977. *Cudonigera*: A new genus for moths formerly assigned to *Choristoneura houstonana* (Tortricidae). J. Lepid. Soc. 31:119-123.

Robinson, C.T. 1869. Notes on American Tortricidae. Trans. Am. Entomol. Soc. 2:261-288.

Sanders, C.J. 1969. Extrusion of the female sex pheromone gland in the eastern spruce budworm, *Choristoneura fumiferana* (Lepidoptera: Tortricidae). Canad. Entomol. 101:760-762.

Sanders, C.J. 1971a. Sex pheromone specificity and taxonomy of budworm moths (*Choristoneura*). Science 171:911-913.

Sanders, C.J. 1971b. Daily activity patterns and sex pheromone specificity as sexual isolating mechanisms in two species of *Choristoneura* (Lepidoptera: Tortricidae). Canad. Entomol. 103:498-502.

Sanders, C.J. 1971c. Laboratory bioassay of the sex pheromone of the female eastern spruce budworm, *Choristoneura fumiferana* (Lepidoptera: Tortricidae). Canad. Entomol. 103:631-637.

Sanders, C.J., G.E. Daterman, R. Shepherd, and H. Cerezke. 1974. Sex attractants for two species of western spruce budworm, *Choristoneura biennis* and *C. viridis* (Lepidoptera: Tortricidae). Canad. Entomol. 106:157-159.

Smith, S.G. 1953. Reproductive isolation and the integrity of two sympatric species of *Choristoneura*. Canad. Entomol. 85:141-152.

Stark, R.W., and J.H. Borden 1965. Life history of *Choristoneura lambertiana subretiniana* Obraztsov (Lepidoptera: Tortricidae) attacking lodgepole pine. Canad. Entomol. 97:684-690.

Stehr, G. 1967. On coniferophagous species of *Choristoneura* (Lepidoptera: Tortricidae) in North America. II. Geographic distribution in accordance with forest regions. Canad. Entomol. 99:456-463.

Swaine, J.M. 1922. The spruce budworm in Quebec Province. *In*: 14th Ann. Rept. Que. Soc. Prot. Plants, 1921-22:32-39.

Swaine, J.M., and F.C. Craighead. 1924. Studies on the spruce budworm (*Cacoecia fumiferana* Clem.). Part I. A general account of the outbreaks, injury, and associated insects. Canad. Dept. Agric. Bull. N.S. 37:1-27.

Walsingham, Lord Th. de Grey. 1879. Illustrations of typical specimens of Lepidoptera Heterocera in the collection of the British Museum. Part IV. North American Tortricidae. British Museum, London; 84 pp. + 17 pls.

Weatherston, J., W. Roelofs, A. Comeau, and C.J. Sanders. 1971. Studies of physiologically active arthropod secretions. X. Sex pheromone of the eastern spruce budworm, *Choristoneura fumiferana* (Lepidoptera: Tortricidae). Canad. Entomol. 103:1741-1747.

2

Biological Relationships

Host Tree Preferences and Isolation by Pheromones Among Allopatric and Sympatric Populations of Western *Choristoneura*

Jerry A. Powell
Essig Museum of Entomology
University of California, Berkeley

John A. De Benedictis
Department of Entomology,
University of California, Davis

Abstract

We investigated biological relationships in natural populations that were at low, non-defoliating levels. We concentrated on regions of sympatry of two or more species, primarily in California but also in Arizona, Nevada, Oregon, and Utah. We documented the constancy of larval hostplant adherence in *Choristoneura retiniana* (Walsingham) and *C. carnana* (Barnes and Busck) by 90 larval collections that yielded adults, in California and Oregon during 1970-1982. Larval host data for *C. occidentalis* Freeman and *C. lambertiana* (Busck) are compared, primarily from specimen and literature records. We used traps baited with virgin females or synthetic pheromones to test attractiveness of males at 28 localities in California, Nevada, Oregon, and Utah, including cross attraction tests with females from distant populations, at 18 sites. We monitored seasonal phenology and tested females from 9 California populations at Blodgett Forest, El Dorado Co., California, where *C. retiniana* and *C. carnana* are sympatric.

Three lineages are defined, adults of which fly synchronously: the *C. retiniana* and *C. carnana* lineages feed on Abietoideae, and the *C. lambertiana* lineage on pines. *C. retiniana* feeds on *Abies*, has unpigmented larval integument and weakly pigmented pupal integument (both appear green in life), and dimorphic color forms in both males and females. Its principal sex pheromone component is E-11 tetradecenyl acetate. *C. carnana* is a lower elevation species that is sympatric with *C. retiniana* at middle elevations in California; it feeds primarily on *Pseudotsuga*, has heavily pigmented larval and pupal integument (appear brick red in life), and adults are monomorphic within populations. The principal component of its pheromone is E-11 tetradecenal. *C. occidentalis* is the northern Cascade Range, Great Basin ranges, and Rocky Mountain counterpart of *C. carnana*, sharing the same pheromone; it feeds on various Abietoideae, primarily *Pseudotsuga*, and has heavily pigmented but variable larval morphs. In most populations the adults are sexually dimorphic, females tending to be rust colored, males dark brown; and the species is geographically polytypic, adults varying in size and forewing pattern from one region to another. *C. lambertiana* feeds primarily on *Pinus*, has heavily pigmented larvae and pupae, and also is markedly polytypic but not sexually dimorphic. Males respond to a synthetic pheromone with the principal components E- and Z-11 tetradecenyl acetate.

In California, Nevada, and Utah, pheromone chemistry is the primary mechanism for maintenance of the species, through intraspecific recognition; interpopulational attraction tests with virgin females showed 100%, and sympatric synthetic pheromone tests showed 99%, separation of males in natural populations. Northward, at the Oregon border (42° N lat.), the reproductive isolation between the two Abietoideae-feeders (and possibly some pine-feeding populations as well) breaks down, and the distinctive adult and larval morphs, hostplant selection, and chemical attraction blend. Populations on the west side of the Cascade Range in Oregon, both sides in Washington, and in the Basin Ranges and Rocky Mountains, are less variable within populations and appear not to be involved in the hybridization. By convention these have been treated as the single composite species, *C. occidentalis*.

INTRODUCTION

We addressed the taxonomic problems outlined in Part 1 (Powell 1995) by investigating biological relationships in natural populations that were at low, non-defoliating levels. We concentrated on regions of sympatry of two or more species in California, and we also sampled in Nevada, Utah, Arizona and Oregon. Populations in the critical region of apparent hybridization in southern Oregon were studied concurrently by W. J. A. Volney and his associates (Volney *et al.* 1983a, b, 1984; Liebhold *et al.* 1984; Liebhold and Volney 1984). Therefore, we did not emphasize that area.

We used three approaches:

1. Surveys using blacklight and/or synthetic pheromone lures to attract adults in localities where budworm taxa were poorly understood or unknown, to document the geographical distributions and ecological associations of populations.
2. Collections of larvae and subsequent rearing to document host tree preferences.
3. Tests of male response in natural populations to synthetic sex pheromones and to virgin females. We reasoned that differences in mate recognition among individuals of sympatric populations act as premating, reproductive isolation mechanisms between species and therefore provide the key to species identification.

METHODS

Nomenclature

The names used in this report follow Powell (1980, 1983). Trinomial designations (e.g., *Choristoneura carnana californica*) are used to specify geographic components of species where names are available, even though application of the subspecies concept in the conifer-feeding *Choristoneura* is subjective. We use binomial designations (e.g., *Choristoneura lambertiana*) to refer to the species in the broad sense, including its described subspecies.

Field Sampling Areas

The primary sites selected for study of biological isolation and seasonality among sympatric species, and for sources of host tree preferences and virgin females used in attraction tests from allopatric populations, were as follows:

Angwin and Cobb areas, Napa and Lake Cos., California: *Choristoneura carnana californica* Powell.

Blodgett Forest, E of Georgetown, El Dorado Co., California: *C. carnana* (Barnes and Busck) (Sierra Nevada population) and *C. retiniana* (Walsingham).

Donner Lake and Tahoe City area, Placer Co., California: *C. retiniana* and *C. lambertiana subretiniana* Obraztsov.

Tehachapi Mtn. Park, Kern Co., California: *C. retiniana.*

Manker Flat, NE of Mt. Baldy P.O., San Bernardino Co., Calif.: *C. carnana carnana* and *C. retiniana.*

Lehman Creek area, Snake Range, and Timber Creek, Schell Creek Range, White Pine Co., Nevada: *C. retiniana* and *C. occidentalis* Freeman.

We also surveyed in numerous areas where budworms previously were not recorded or were known from isolated specimen records, as described below (Fig. 2.1, Appendix 2.1) (De Benedictis Part 4 this vol., Appendix 4.1.)

Techniques

Rearing Field Collected Larvae

All areas that we surveyed contained low density populations (0-10 larvae/search hour), with the exception of Tehachapi Mountain, which had moderately dense populations in 1980-1982 (40-50 larvae/search hour). We visually searched subcanopy foliage, especially along margins of open areas, and sampled by using beating sheets at 1-2 m heights. We employed pole pruners to sample at heights of 3-6 m, but random surveys by pole pruner proved negative in 1979 at all sites and were not attempted thereafter. In the field, larvae were placed in polyethylene bags with foliage and transported in ice chests. In the laboratory, larvae were segregated into small lots (n=1-6) in polyethylene bags or translucent plastic rearing boxes and reared on foliage, with daily surveillance to regulate moisture conditions. Each collection of one or more larvae is identified by a lot number based on the year and month of collection: e.g., 80F12 was the twelfth collection in June 1980.

Testing Mate Acceptance and Oviposition in Laboratory Conditions

Male-female pairs of lab-reared, virgin moths were placed in small cages made from one-pint cardboard food containers with a ceiling of nylon netting held in place by a plastic snap-top rim. Each cage was provided with foliage of the appropriate host tree, held in a cotton-stoppered water vial that also served as a moisture source for the moths.

The containers were placed on shelves adjacent to open windows to enhance air circulation, and the moths were observed periodically during evenings for mating activity. Females that did not produce viable eggs were later dissected to ascertain presence or absence of spermatophores.

Trapping Males by Attraction to Sex Pheromones

Trap methods

We used synthetic pheromone lures and virgin females to attract male moths in natural populations. Females that had been lab reared from larvae were housed in nylon screen-sided cages 3.7 x 3.7 x 3.0 cm in outside dimensions, which were constructed from cardboard pill boxes, or in similar sized aluminum screen cages. Each was provided with water in a cotton-stoppered vial. After transport to the field site in an ice chest, the caged female or synthetic bait was suspended within a Zoecon Pherocon II sticky trap. In 1981 we also used commercial antifreeze as a trapping agent at some localities, in traps constructed from plastic refrigerator boxes (ca. 10 x 10 x 7 cm deep, with windows cut in each side). Experiments at Blodgett Forest were inconclusive as to whether these or sticky traps were more effective. The liquid traps have practical advantages in that suitable study specimens can be removed easily when daily monitoring is conducted, and the traps are reusable.

When traps were deployed for several days or longer, there are disadvantages to both kinds of traps. Specimens become soaked in the antifreeze, which bleaches the orange colors and makes assessment of the phenotype difficult. Also, the contents can be spilled by birds or small mammals. Specimens in sticky traps damaged themselves or were damaged in removal so that most were unsuitable for study. In moderately dense populations, the sticky surfaces become saturated with moths and debris, so that after one or a few nights the catching efficiency is reduced. In these circumstances scavengers, especially vespid wasps and ants, often damage the catch. Therefore during season-long surveys, such as at Blodgett Forest, we replaced the sticky traps weekly.

Traps baited either with females or with synthetic pheromone baits were deployed 20 or 30 m apart, hung from branches 1.5-2.5 m above the ground. In 1979 trap lines were deployed alternating matching and non-matching host trees of the respective baits. There were no statistically significant differences in attraction of *Choristoneura carnana* with respect to tree species, so during subsequent years individual trap sites were selected for suspected air movement characteristics rather than tree species.

Each trap line consisted of 3 or 4 replications of each pheromone bait, and an unbaited trap. No *Choristoneura* were trapped in any of more than 50 unbaited controls, and therefore any male taken in the baited traps was interpreted as having been attracted to the chemical.

Attraction by virgin females

Using virgin female-baited traps, a pilot test with *C. carnana californica* at Angwin in 1980 demonstrated that calling activity of females could be delayed 6-10 days by refrigeration (4 $\pm$ 1.5°C) without appreciable loss of attractiveness to males. We delayed emergences of reared adults 7-10 days by chilling pupae and/or delayed the activity of newly emerged females 6-8 days by refrigeration, but numbers of males trapped diminished when female or pupa + adult chilling exceeded 10 days (Table 2.1). Therefore, pupae and/or newly emerged females were refrigerated to regulate contemporary numbers for mating trials, and to synchronize the availability of virgin females for field tests of pheromone attraction with the natural flight period of males from other localities.

Among females of *Choristoneura carnana californica* deployed at Blodgett Forest, 90% survived more than 7 days in the traps. Early in the season, most lived more than 15 days. In 1980, males (n=209) were attracted mostly during the first 7 days after females were deployed, but several were trapped between days 8-12, and 2 were attracted as late as 17-21 days by females deployed early in the season. No females of *C. retiniana* were available for early season tests, and those deployed in midseason averaged shorter longevity than *C. carnana*, but 70% survived more than 7 days. Only two *C. retiniana* males were attracted after 7 days. The traps for both species were monitored after one night and weekly thereafter, so daily attraction of males was not tabulated.

At other stations, virgin females or synthetic lures were deployed for single nights, or traps were left for retrieval after several days or weeks.

In 1980-1982, virgin females were obtained by larval rearing and deployed for interpopulational attraction tests as summarized in Table 2.2 and Fig. 2.2. The primary goal of these experiments was to confirm that morphologically similar populations in distant localities produce and respond to consistent sex pheromones. The rationale for bringing females to the central Sierra Nevada was to test attractiveness of pheromones in populations of *Choristoneura retiniana* and *C. carnana* that are allopatric from one another to response by males in a region of sympatry of the two species. Blodgett Forest, at 1,300 m elevation, is situated in mixed *Abies* and *Pseudotsuga* that supports the sympatry. Blodgett is a University of California station located three hours drive from Berkeley, so it provided a convenient site for weekly monitoring of the traps, wherethey were protected from potential vandalism.

Synthetic pheromone lures

The synthetic lures were provided by G. E. Daterman, who carried out research on the nature of the sex pheromones and development of the baits concurrently with our studies (Daterman et al. 1983, 1984, Part 3 this vol.). The formulation that had been developed for *C. fumiferana* and found to be effective for *C. occidentalis* in the Pacific Northwest (Weatherston *et al.* 1971), attracted males of *C. carnana* in California and *C. occidentalis* in Nevada and Utah from the onset of our study. The baits used for *C. retiniana* and

C. lambertiana became increasingly effective during 1980-1982 as more refined blends were developed.

The names we use to designate the synthetic lures and their chemical formations are listed in Table 2.3. We tested a series of baits based on the pheromone of *C. retiniana* in Modoc Co., California. We refer to both the 1978 and 1979 lures as RET-A bait, as both failed to attract any *Choristoneura* males in California, Nevada, and Utah, although they drew several non-target species consistently. In 1980 a few *C. retiniana* came to the RET-B bait, and that was succeeded by effective lures designated RET-C in 1981 and 1982.

Two baits based on the pheromone of *C. lambertiana subretiniana* in Oregon were tested from 1980 to 1982. The SUBRET-A lure was ineffective, but it was succeeded by the biologically active SUBRET-B baits used in 1981 and 1982 (Table 2.3).

The effective lures attracted males of the three conifer-feeding species groups at either 0.1% or 0.01% active concentration, despite their different release rates.

Survey Trapping with Blacklight

Blacklight attraction was used in conjunction with synthetic pheromone lures in geographic areas that were represented by too few specimens to enable adequate taxonomic assessment and/or where host tree associations were unknown. We selected sites by suitable potential host tree composition and set up two vertical sheets with 15-watt blacklights powered by 12-volt batteries. Moths attracted to the sheets were collected between approximately 2100 and 2400 hrs, and all night by blacklight traps. Fresh appearing females of *Choristoneura* were retained alive for oviposition.

Field Survey

1978

We conducted a pilot survey in late July-August in El Dorado Co., California, employing blacklight and traps baited with OXY lure and RET-A lure. We trapped at four sites along an elevational transect from 1,300 m at Blodgett Forest, in mixed *Abies-Pseudotsuga* forest, to 2,000 m at Loon Lake, which is above the limit of *Pseudotsuga* at this latitude. Both *Choristoneura retiniana* and *C. carnana* were attracted to lights at Blodgett Forest, but males of only the latter were trapped by the OXY lure (n=70). The RET-A bait was ineffective. The results indicated that there are chemical differences in sex attractants and suggested that *C. carnana* is associated with *Pseudotsuga*, because it appeared at all sites except Loon Lake.

In addition, the same baits were used during 1-2 night tests at three sites in eastern Nevada and the Wasatch Range, Utah, with similar results—positive for *C. occidentalis*, negative for *C. retiniana* —although both species were taken at blacklight in Nevada.

1979

We carried out larval surveys (83 collections) in central California, in the Coast Ranges (Lake, Napa, and Sonoma Cos.) and in the Sierra Nevada (Blodgett Forest and Lake Tahoe-Truckee area). The results showed that *C. carnana* lives on *Pseudotsuga* in the Coast Ranges and Sierra Nevada within and below the elevation of *Abies*, and that *C. retiniana* feeds on *Abies* in the Sierra Nevada and North Coast Ranges and extends well above the zone of *Pseudotsuga*.

We extended the geographical survey with blacklight and/or the same synthetic pheromones as used in 1978 to the Coast Ranges, including the Santa Cruz Mts. in San Mateo Co., Sierra Nevada, Tehachapi Mts., and San Gabriel Mts. Trapping was carried out at 10 locations, from 1-34 nights (mostly 18-34).

1980

Larval collections in Lake Co. (*Choristoneura carnana californica*), and in the Tehachapi and Warner Mts. and Lake Tahoe area (*C. retiniana*) were made primarily to rear virgin females for interpopulational attraction tests. We also extended the larval survey to the Klamath-Salmon Mts. and southern Cascade Range of northwestern California (Shasta-Trinity region), and to the San Gabriel Mts., to document the host plant preferences of *C. carnana*. Some females from these areas were used in interpopulational attraction experiments (Table 2.2).

We monitored the seasonal flight and differential attraction of males by traps baited with synthetic pheromone lures and virgin females throughout the season at Blodgett Forest and surveyed the pheromone responses of sympatric species at several localities in northeastern California and eastern Nevada. The RET-B lure was only weakly effective, but virgin females deployed at the same sites provided supporting data. Tests of attraction by virgin females of males in distant populations were conducted as indicated in Table 2.2 and Figure 2.2.

1981

Larval collections were made only in Napa Co. (*Choristoneura carnana californica*) to provide livestock for use by Volney in mating trials and by Daterman for pheromone analysis, and at Tehachapi Mountain (*C. retiniana*) to provide material to Stock for isozyme analysis and virgin females for our interpopulational attraction tests.

Early season trials of RET-C lure at Blodgett Forest were effective. Therefore we expanded survey efforts to obtain attraction data in Nevada, Arizona, and Utah. A revised formulation, SUBRET-B bait, also proved effective for the pine-feeding populations.

We used the three baits to survey four sites along the east side of the northern Sierra Nevada during 28- to 44-day periods. We trapped with blacklight and synthetic pheromones for 1-2 nights, at 12 sites in the Great Basin Ranges: Spring Mts. in southern

Nevada, Flagstaff area of northern Arizona, Cedar Breaks area and Wasatch Range, Utah, and Snake Range and Copper-Jarbidge Mts. in eastern and northern Nevada.

1982

We used virgin females of *Choristoneura retiniana* from Tehachapi Mountain and of *C. occidentalis* from the USFS lab colony in Berkeley, which originated from northern Idaho-Montana (Lyon *et al.* 1972), along with synthetic baits, to survey population composition in northern California and Oregon (Fig. 2.3). Selecting sites shown by Volney's group and our 1980 larval collections to harbor taxonomically confusing populations, we surveyed 14 localities in a broad arc from the Klamath Ranges of Siskiyou Co. northward to east-central Oregon, for 1- to 8-night periods.

We monitored the flight periods of *C. carnana* and *C. retiniana* at Blodgett Forest by continuous pheromone bait trapping and weekly blacklight sampling. We also surveyed four localities in the northern Sierra Nevada (Plumas Co.) with the three pheromone lures, during 44- to 90-day periods.

1988

We made collections of larvae from *Abies* and *Pseudotsuga* at several sites along the eastern slope of the Cascade Range and Simcoe Mts. in Washington (Okanogan, Kittitas, Yakima, Klickitat Cos.). The goals were to determine if populations north of the Columbia River maintain heterogeneous, hybridizing characteristics, and to provide virgin females for tests of attractiveness to *Choristoneura carnana* and *retiniana* in the Sierra Nevada (De Benedictis *et al.* Part 5 this vol.).

RESULTS AND SIGNIFICANCE

Geographical Survey

A major problem hindering early taxonomic assessments was lack of adequate population samples. Available material in collections consisted of geographically scattered samples, often of one or a few specimens, and most lacked hostplant data (Freeman 1958, Obraztsov 1962, Powell 1964). Later analyses were based on laboratory colonies originating from areas of outbreak populations (Freeman 1967). Therefore, one goal of this project was to survey by synthetic pheromone trapping and larval collections as comprehensively as possible, emphasizing known or suspected regions of sympatry among two or more species.

Figures 2.1 and 2.2 depict localities surveyed during 1978-1982, either by larval collection, blacklight, or pheromone trapping, or comprehensively, employing all three approaches. Many of these populations previously were known from only one or a few specimens. Discoveries include the following components of geographic distributions:

Choristoneura retiniana

The northern Sierra Nevada (Plumas Co.) at five sites, and in the Transverse Ranges of southern California (Mt. Pinos, Ventura/Kern Co.; San Gabriel and San Bernardino Mts.), and the Great Basin Ranges of eastern Nevada (Schell Creek and Snake Ranges and Charleston Mtn.) and Utah (Wasatch Range and Cedar Breaks area).

Recovery of *C. retiniana* specimens in Utah that show the same dimorphism as is displayed by Sierra Nevada, Tehachapi, San Gabriel and Spring Mts. populations has nomenclatural significance. It confirmed our suspicion that *C. spaldingiana* (Obraztsov 1962), originally described from one worn specimen collected in 1909 from Provo, Utah, is the mottled forewing morph of those populations of *C. retiniana*. Obraztsov noticed the resemblance of the type specimen to mottled forms from Lake Tahoe, but he failed to recognize the relationship of the dimorphic color forms. He named the tan form *lindseyana*, including material from both the Sierra Nevada and Modoc Co. Our collections from Mt. Timpanogos near Provo (n=33), taken at RET-C bait, included 20 of the mottled form that match the type specimen .

Choristoneura carnana

Coast Ranges of central (Santa Cruz Mts.) and northern California, Trinity-Salmon Mts., and southern Cascade Ranges (*C. carnana californica*); northern Sierra Nevada; and mountains of southern California (*C. carnana carnana*).

Choristoneura occidentalis

Northern Nevada (Copper-Jarbidge Mts.) and eastern Nevada (Schell Creek and Snake Ranges); Wasatch Range and southern Utah.

Choristoneura lambertiana

Northern Sierra Nevada and San Bernardino Mts., California (*C. lambertiana subretiniana*); Copper-Jarbidge and Charleston Mts., Nevada; Flagstaff area, Arizona; southern Utah (*C.lambertiana near ponderosana*).

Confirmed sympatry of two or more species include the following:

1. *C. retiniana* and *C. carnana* in California where *Abies* and *Pseudotsuga* occur together: the San Gabriel and San Bernardino Mts., west slope of the Sierra Nevada at middle elevations, Plumas County east of the sierran crest, and at Mt. Shasta.

2. *C. retiniana* and *C. lambertiana subretiniana* at higher elevations in Plumas Co., southward to Lake Tahoe, California.

3. Coexistence of three species, *C. retiniana spaldingiana*, *C. l. ponderosana,* and *C. occidentalis*, in various combinations, in the Great Basin Ranges:

C. r. spaldingiana and *C. l. ponderosana,* but not *C. occidentalis,* at Charleston Mountain, southern Nevada;

C. occidentalis and *C. l. ponderosana*, but not *C. r. spaldingiana*, in the Copper-Jarbidge Mts., northern Nevada, and near Flagstaff, northern Arizona;

C. retiniana and *C. occidentalis*, but not the pine-feeding species, in the Schell Creek and Snake Ranges, eastern Nevada, and in the Wasatch Range, Utah.

Sympatry of all three species in Gold Lake area, Plumas Co., California, and in Cedar Breaks area, Kane Co., southern Utah.

Larval Host Tree Associations

During 1979-1982 we made about 240 collections of foliage-feeding larvae from Pinaceae in widespread areas of California and Oregon (Fig. 2.1). Adults of *Choristoneura* were reared from 66 collections, including 33 of *C. carnana*, 16 of *C. retiniana*, 13 of hybrid populations, and 4 of *C. lambertiana subretiniana.* These data and those from collections made during 1970-1974 are summarized in Table 2.4 and Appendix 2.1.

Larval, pheromone trap, and blacklight collections indicate that throughout its range *Choristoneura retiniana* occurs in association with *Abies*. This species has been reared from *Abies* at 25 localities in California (Appendix 2.1), and we have no records on other conifers. *C. retiniana* lives primarily in association with *A. concolor* in California and the Great Basin ranges and feeds on *A. magnifica* at higher elevations in the Sierra Nevada and on *A. grandis* or hybrids in southern Oregon. Populations are not known in association with the coastal *A. grandis* in California.

Most of the rearing records of *C. retiniana* are from elevations above, or in mountain ranges outside, the distributions of *Pseudotsuga* and *C. carnana.* Collections from areas of sympatry (McBride Springs and Snowmans Summit at Mt. Shasta; Blodgett Forest) involved few individuals.

The larvae and pupae of *C. retiniana* have unpigmented integument, appearing bright green in life (hence the common name "Green Budworm"), and bear strong cryptic resemblance to the new fir foliage.

Choristoneura carnana feeds primarily on *Pseudotsuga* (16 localities, 28 collections, Appendix 2.1). In southern California, we confirmed that *C. carnana carnana* feeds on *P. macrocarpa*, and its higher elevation colonies are sympatric with *C. retiniana*, where *Abies* occurs. Populations of *C. carnana* in the Sierra Nevada are presumed to feed on *P. menziesii*. Several small collections were made at Blodgett Forest (n=10 reared), and one *C. carnana* was reared from a larva found on *Abies concolor* in a mixed tree stand. All attempts to trap with OXY lure at sites higher in elevation than *Pseudotsuga* occurs, in the northern and central Sierra Nevada (Bucks Lake, Plumas Co.; Donner Lake-Lake Tahoe area; Loon Lake, El Dorado Co.) attracted no *Choristoneura*, indicating that *C. carnana* does not maintain populations at elevations above the occurrence of *Pseudotsuga*.

Choristoneura carnana californica feeds on *P. menziesii* throughout the North Coast Ranges. In the southern Klamath Ranges, larvae were found on *Pseudotsuga* and *Abies concolor* at three sites (Appendix 2.1). At Blue Mountain and Nelson Creek Gap, 14 larvae were taken from *Abies* (26% of the sample). Most of these were fully grown larvae which were discovered, conspicuous by their contrasting brick-red color, on fir saplings that showed little evidence of feeding. These larvae may have descended from the mixed fir and Douglas-fir canopy. A few, however, were located in feeding shelters on lower limbs of larger *Abies*. Although *C. retiniana* was not found with this population, coexistence of the two *Choristoneura* has been detected a short distance to the north.

Thus it appears that *C. carnana* depends primarily upon *Pseudotsuga* and does not occur outside the range of that genus. In mixed stands of *Pseudotsuga* and *Abies*, such as in Trinity, southern Siskiyou, and El Dorado Cos., it feeds on both, at least in later instars (Appendix 2.1).

Larvae and pupae of *C. carnana* are heavily pigmented with reddish brown. Early instar larvae are cryptically concealed along newly elongating shoots, which are reddish, and in shelters that incorporate pieces of the recently opened, red-brown bracts. Later the larvae feed in shelters constructed of drawn-together maturing needles, which are darker green than those of *Abies concolor*.

The geographical distribution of *Choristoneura lambertiana subretiniana* corresponds with that of Lodgepole Pine (*Pinus contorta murrayana*) in California, and larvae have been reared from that pine several times in the Truckee area and Warner Mts., Modoc Co. (Powell 1964; Stark and Borden 1965; NMNH specimens; E. A. Cameron unpubl., in litt.; present data). Similar forms have been reared from *Pinus ponderosa* at three localities in eastern Oregon (Hopkins U.S. data). However, there are also records from *Abies* and other conifers (Table 2.4, Appendix 2.1). Cameron (in litt.) surveyed budworm in the area 12 miles north of Truckee in July 1964, where *C. l. subretiniana* and *C. retiniana,* but not *C. carnana,* coexist. He found larval abundance to be about 100/search hr on Lodgepole Pine, 4/hr on *P. jeffreyi*, and 25/hr on *A. concolor*, and "only a few of the green form [*C. retiniana*] were encountered." Evidently *C. l. subretiniana* is dependent primarily upon pine, and is geographically distributed accordingly, but occasionally feeds on other Pinaceae, at least in later instars in mixed tree stands.

Biological Isolation Between Sympatric Species

Reproductive Compatibility in Laboratory Conditions

Prior studies have shown that *Choristoneura* species that are reproductively isolated in natural situations can interbreed in the laboratory (Smith 1953, Campbell 1967). Isolation between two sympatric species in eastern North America, *C. fumiferana* (Clemens) and *C. pinus* Freeman, is maintained in the field by interaction of ecological, temporal and physiological-behavioral factors (Smith 1953, Sanders 1971). Larvae of *C. fumiferana* feed primarily on spruce and fir, while those of *C. pinus* feed almost exclusively on pines;

females mate near the pupation site and males tend to restrict activities to the host trees. Adults of *C. fumiferana* emerge about 10-14 days ahead of sympatric *C. pinus*, so that males of the former are mostly gone by the time female *C. pinus* emerge. Temporal separation is enhanced by differences in daily activity: female *C. fumiferana* begin accepting males earlier in the day, with 50% of the females active by sunset, and calling activity peaks 2.5 hours later; by contrast, 20% of *C. pinus* females begin calling by sunset, and calling activity peaks about 5 hours later.

Smith (1953) reported that when equal numbers of males and females from sympatric populations of these two species were placed in the same cages, about 18% of the observed matings were interspecific. He found no exceptional sterility in the F_1 hybrids. Some F_1 females showed variable sterility that Smith attributed to non- homologous chromosomal configurations seen in their brothers, but the sterility was inconsistent and difficult to assess. Such sterility did not appear in subsequent crosses of hybrid females between the two eastern species, nor in crosses with the then unnamed *C. occidentalis* from the Rocky Mts..

Campbell (1967) tested mate choice between allopatric pairs of *Choristoneura: C. fumiferana* and two species from western Canada, *C. occidentalis* and *C. biennis* Freeman. He found that 85% of *C. fumiferana* females mated conspecifically, while those of the other two species were indiscriminate. The studies indicate that *Choristoneura* are less capable of maintaining specific isolation when confined in lab tests than under natural conditions. Such mate-choice tests may greatly underestimate the natural reproductive isolation that exists between sympatric populations. Sanders (1971) suggests that normal behavioral limitations break down because males moving around in the restricted space contact females by chance, and having done so, may mate with females of either species. Budworm researchers have concluded that this capability applies to any pair of *Choristoneura* species, although most western populations have not been tested.

We found that *C. retiniana* and *C. carnana* can be reproductively compatible in laboratory conditions. We did not obtain sufficient numbers of specimens from sympatric populations, so the trial crosses were from allopatric populations in which *C. carnana* emerges earlier in the season; hence only older *carnana* males were available. Two sets of four pairs were tested in July 1980. In the first group, 22- to 28-day old refrigerated males of *C. carnana* (80F12, Lake Co.) were confined with 4- to 6-day old females of*C. retiniana* (80F25, Tehachapi Mt.). In the reciprocal tests, males of *C. retiniana* that were 2 to 6 days old (80F105, Modoc Co.) and 3- to 8-day old females of *C. carnana* from the Trinity-Shasta region (80F43, F58, F70, F73) were paired. The older males of *C. carnana* lived only 2 to 4 days, females of *C. retiniana* 5 to 10 days; the males of *C. retiniana* survived only 3 to 5 days and females of *C. carnana* 6 to 10 days.

Three of the *C. retiniana* did not oviposit, and the fourth laid only a few infertile eggs. Dissections revealed that none of the females carried spermatophores, indicating that mating was not culminated. The age of the males may have been the primary deterrent. All of the *C. carnana* females produced eggs, but those of three were deposited in irregular patches or singly and were infertile. The fourth female produced 105 eggs, mostly in regularly imbricate masses; 76% developed to eclosion, and the first instar larvae constructed hibernacula normally. Dissections showed that one of the three remaining females carried a spermatophore; she had produced about 70 eggs that failed to develop, suggesting post-mating, embryonic sterility in some F_1 hybrids.

In control tests of laboratory mating, males and females from the Angwin population of *Choristoneura carnana californica* were confined in June 1980. Eight reared, virgin females that were 4 to 20 days old were paired with 4 reared, 6- to 15-day old males and 4 males taken at blacklight 4 days prior to the trial. Six of the 8 pairs produced eggs, of which 4 families were viable, half of them fertilized by field collected males, half by reared males. The results indicated that moths of various ages confined in the conditions we provided would not be expected to produce 100% successful matings, even if there were no barriers to mate acceptance or reproductive compatibility.

The fact that one of four female *C. carnana* X male *C. retiniana* pairs mated and produced viable eggs in confinement suggests that hybridization between these two species could occur in crowded conditions (i.e., hybridization could begin as a result of epidemic population levels of one or both species).

Reproductive Isolation in Natural Populations

Evidently reproductive isolation between *Choristoneura retiniana* and *C. carnana* in areas where they coexist in California is maintained by ecological and physiological-behavioral mechanisms. We investigated three lines of evidence relevant to this hypothesis: host tree associations, seasonal flight periods, and attraction to species-specific sex pheromones, employing synthetic baits and virgin females.

Host tree associations

The larval host tree data in California (Table 2.4, Appendix 2.1) indicate that each species has a preferred conifer upon which it primarily depends: *Choristoneura retiniana* on *Abies, C. carnana* on *Pseudotsuga,* and *C. lambertiana* on *Pinus*. In mixed tree stands, however, there is some crossover to other conifers, at least by *C. carnana* and *C. l. subretiniana*. Such exchange may occur only after dispersal by post-diapause larvae, rather than by oviposition preference. It appears that host selection enhances the integrity of species by partitioning emergence and mating sites, but probably it is not a primary isolating mechanism between sympatric populations.

Seasonal flight periods

Two species of conifer-feeding *Choristoneura* often have been observed together, either as adults or larvae, but activity of sympatric species discussed here has been monitored throughout the season at only a few sites in California. At Blodgett Forest, flight periods of *C. retiniana* and *C. carnana* were documented in 1980, 1981, and 1982. Activity of *C. retiniana* and *C. lambertiana subretiniana* was recorded by light trap collections at the Warner Mts. during five weeks in 1922 by A. W. Lindsey (Powell 1964:176, 183) and throughout the season at Tahoe City, Placer Co., during 1974-1982 by N. Westerland.

In each instance, the spans of seasonal activity and the peak abundance of both species are approximately the same (Figs. 2.4, 2.5). Our data at Blodgett Forest originate from continuous trapping of males attracted to synthetic pheromones and/or virgin females from distant populations, supplemented by weekly blacklight collections. The 1980 data are not fully comparable between species because the RET-B bait was only weakly effective, and because virgin females of *C. retiniana* were not available until 14 days after the study began, while *C. carnana* females were deployed from the beginning. The 1981 data are incomplete because sampling was initiated after the flight period had begun. Nonetheless, the pattern of activity, as represented by cumulative percent of total catch, was the same as in 1982, when the data were more complete.

At Blodgett Forest, the earliest *Choristoneura carnana* preceded the first *C. retiniana* by 7-10 days, roughly corresponding to the differences in foliation phenology of their respective primary host trees; but peak numbers, as indicated by 50% of the individuals, were reached by *C. retiniana* only two days later than *C. carnana* (Fig. 2.4). The two were active together during most of the flight season. Comparable data for *C. retiniana* and *C. lambertiana subretiniana*, at Lake Tahoe, where Westerland sampled with blacklight, reveal the same relationship (Fig. 2.5). The two species fly synchronously, with *C. l. subretiniana* slightly earlier in 1974, and *C. retiniana* about five days earlier at peak flight in 1981, the years for which the most complete records are available.

The data indicate that seasonal displacement is not a factor in reproductive isolation of sympatric species among *C. retiniana*, *C. carnana* and *C. l. subretiniana.* Collections of 1-2 nights in Nevada and Utah suggest that the same relationship obtains between sympatric *C. retiniana, C. occidentalis,* and *C. lambertiana.*

Attraction to species-specific sex pheromones

Synthetic pheromone baits

Evidence for the biological isolation of *Choristoneura* species was provided in two ways by use of synthetic pheromones: (a) differential attraction of males at sites where two or more species are sympatric; and (b) attraction at sites characterized by differing potential host tree species. The data provide evidence that supports interpretation of host tree specificity, as described above.

Formulations approximating the sex pheromones of the three species-groups were employed (Table 2.3). While the OXY lure was effective from the onset of the study, the RET baits in 1978-1980 and SUBRET baits were effective beginning in 1981.

Differential attraction in sympatric populations

Choristoneura retiniana and *C. carnana*. Trapping was carried out at Blodgett Forest during five seasons, 1978-1982. There was 100% separation of males attracted to the baits. The RET-A bait used in the first two seasons was ineffective, but revised formulations resulted in capture of 19 *C. retiniana* in 1980 (RET-B bait) and 391 in 1981-82 (RET-C bait) (Table 2.5).

Sympatric populations of the same two species also differentiated between the two baits along the eastern edge of the Sierra Nevada in Plumas County (Table 2.6). Traps with RET-C bait averaged 8.7 males of *C. retiniana* (n=157), those baited with OXY lure captured 13.2 males/trap of *C. carnana* (n=252), while neither attracted the reciprocal species, at the four sites characterized by mixed *Abies-Pseudotsuga* forests. Bucks Lake, the northernmost station of the transect, at 1,590 m, and Lake Tahoe, the southernmost site, at 1,900 m, are above the limit of *Pseudotsuga* at those latitudes, and no *C. carnana* were taken.

Choristoneura retiniana and *C. occidentalis*. Isolation of this pair was tested in five mountain ranges of eastern Nevada and Utah (Table 2.7). The RET-B bait used in 1980 was ineffective, but virgin females used concurrently helped confirm the specificity shown by synthetic lure in 1981. The catches of *C. occidentalis* ranged from 0.3 males/trap in one night late in the season (when females predominated at blacklight) at Lehman Creek, to 49 males/trap in 12 nights at Timber Creek and averaged 8.0 males/trap (n=210) for the 26 trap-nights. Fourteen traps with female *C. retiniana* or RET-C lure averaged 4.7 males/trap (n=123) at the same sites, with no reciprocal attraction (Table 2.7). One male of *C. occidentalis* was taken in a RET-B trap in 1980; this bait was ineffective for *C. retiniana* in these populations.

Choristoneura retiniana and *C. lambertiana*. The SUBRET-B bait developed for the pine-feeding complex was effective in 1981-82. Tests of attractiveness were carried out in Plumas Co. and at Lake Tahoe, California, near the site where Westerland accumulated flight period data during several years for *C. retiniana* and *C. l. subretiniana* (Fig. 2.5, Table 2.6). Single night tests were made at Lee Canyon, Charleston Mt., Nevada, and in the Willis Creek-Duck Creek area near Cedar Breaks in southern Utah (Table 2.7), where *C. retiniana* is sympatric with *C. lambertiana* (population morphologically similar to *ponderosana*). For *C. l. subretiniana* (Sierra Nevada), there was a 99% separation, with RET-C baits attracting 97 *C. retiniana*, no *C. l. subretiniana*, while SUBRET traps caught 70 of the latter and one *C. retiniana*, the last at a locality where *C. l. subretiniana* is not known to occur. Males in the Great Basin Ranges discriminated completely between baits, with 14.1 *C. retiniana* males/trap and 5.4 *C. lambertiana* (*ponderosana*-like) males/trap in 7 trap nights of each of the respective baits.

Choristoneura occidentalis and *C. lambertiana*. The integrity of pheromone specificity for *C. occidentalis* and sympatric pine-feeding populations was tested during 1-2 nights in 3 areas: Copper-Jarbidge Mts., northeastern Nevada (3 sites); Willis Creek-Duck Cr. area, southern Utah (2 sites); and San Francisco Mts. area, Arizona (2 sites). In the northern Nevada population there was evidence of incomplete separation, but few *C. lambertiana* were taken (one male at SUBRET-B bait and one in a blacklight trap). About 97.5% of *C. occidentalis* (n=164) were taken at OXY-baited traps, but 3 were attracted to SUBRET-B and one to RET-C, in 12 trap nights of each bait. In the Utah and Arizona populations there was total differential attraction (Table 2.7). Near Flagstaff, Arizona, at Hockderffer Hill, ca 2,500 m, and Mt. Elden, ca 2,650 m, 38 *C. occidentalis* and 34 *C. l.* near *ponderosana* were taken at their respective baits, none reciprocally, in 9 trap nights of each. *C. retiniana* is not known in these Nevada and Arizona areas, although *Abies* and

Pseudotsuga both occur at the Arizona sites. At the Hockderffer Hill site and in the Copper-Jarbidge Mts., there were two discrete size classes of *C. occidentalis* at blacklight (UCB, NMNH specimens), but not at pheromone bait, so a more complex situation than previously supposed may obtain in some populations of *C. occidentalis.*

Interpopulational attractiveness of virgin females

We tested the attraction of males to females imported from distant populations (Figs. 2.2, 2.3). Experiments were carried out with *Choristoneura carnana* and *C. retiniana* females (Table 2.2). Both were deployed at Blodgett Forest, where natural populations of the two species are sympatric and adult activity is synchronous. Both also were tested in Modoc Co., where neither *Pseudotsuga* nor any *C. carnana-occidentalis* series representative was known. *C. retiniana* females alone were deployed at two sites that lack *Pseudotsuga* (Bucks Lake, Plumas Co.; Lake Tahoe), as well as at Sulphur Creek, Plumas Co., where both *C. retiniana* and *C. carnana* occur, and in eastern Nevada where *retiniana* and *C. occidentalis* are sympatric (Fig. 2.2).

All males captured were attracted by their conspecific females (n= 103 *C. retiniana*, 209 *C. carnana*), corroborating results based on synthetic pheromones at the same localities (Tables 2.5, 2.6, 2.7). At Rush Creek, Modoc Co., neither baits nor 3 females of *C. retiniana* attracted any *Choristoneura*, although larval searches had revealed a low density population there. At Cottonwood Springs, Modoc Co., traps baited with 4 *C. retiniana* females caught 33 males, the RET-B bait and one *C. carnana californica* female attracted none, while one *C. retiniana* male was trapped using OXY bait.

The 27 females of *C. carnana* deployed at Blodgett Forest averaged 7.7 males trapped (Table 2.5), while 28 *C. retiniana* females imported to various locations (Table 2.2) captured 5.5 males/trap. Low performance at Bucks Lake and Lake Tahoe, 10 males in 6 traps (Table 2.6), almost certainly was the result of premature deployment, ahead of the main flight period.

Females were tested in low numbers, the number of nights that each attracted males is unknown, and performance of resident females was not matched. Therefore statistical comparisons of interpopulational attractiveness within species are inappropriate. In tests with 10 or more females at Blodgett Forest (Table 2.5), *C. carnana californica* from the Trinity-Siskiyou area attracted 13.5 males/trap, and those from the Lake-Mendocino area captured 4.3 males/trap, while 3 *C. carnana carnana* females attracted 7.0 males/trap.

The greatest distances separating populations used in 1980 cross-attraction tests were San Gabriel Mts. *C. carnana carnana* to Blodgett Forest (580 airline km), and Modoc Co. *C. retiniana* to the Schell Creek and Snake Ranges, Nevada (565-595 airline km). The latter's single trap-night catches of 10 and 14 *retiniana* suggested that these distances have no appreciable effect on pheromone relationships.

Hybridizing Populations in Oregon

In the studies summarized above, there are three series of species that segregate by host tree preference and sex pheromone chemistry in the Sierra Nevada and Transverse Range of southern California, Great Basin ranges, and northern Arizona. Moths of sympatric populations can be distinguished by morphological features (De Benedictis Part 4 this vol.) and by facies, and the larvae and pupae of *Choristoneura carnana-occidentalis* and *C. retiniana* are clearly distinctive. Populations referred to *C. occidentalis* replace *C. carnana* in the Great Basin, where in some areas they are sympatric with *C. retiniana*, while other populations are allopatric, such as in the Copper-Jarbidge Mts., where *Abies* is the only available Abietoid host.

Northward from the Sierra Nevada, however, the situation is not clear cut. Populations of *C. retiniana* in the Modoc-Mt. Shasta region, although they retain the green larval form and appear to maintain reproductive isolation from sympatric *C. lambertiana subretiniana*, display sexual dichromatism and are much more variable in wing color pattern and other morphological details than is true in populations to the south and east (De Benedictis, Part 4 this vol.). The tan form is less numerous, and males are ca. 80% gray-brown or intermediate, while females are all rust- to dark reddish brown (see Figs. 7.11-7.16 and 7.21-7.28 in Powell and De Benedictis, Part 7 this vol.).

In the Mt. Shasta area and westward in the Salmon and Trinity Mts., rearing larvae or pheromone trapping in mixed *Abies-Pseudotsuga* forest produces two separable entities, though not as distinctive as in the Sierra Nevada. *Choristoneura carnana* has lost its distinctive, broadly banded forewing pattern and its peach colored hindwings, to become a variable, usually more reticulate form with gray hindwings (*ibid.*: Figs. 7.47-7.50). In western Siskiyou Co. (Salmon Mts.) the tan phenotype of *C. retiniana* is rare or lacking, and some of the mottled individuals display variable forewing banding resembling that of *C. carnana* (*ibid.*: Figs. 7.17-7.20). There is not a significant difference in size or forewing breadth, as there is in the Sierra Nevada. Other morphological features, such as uncus shape, which is relatively distinctive for populations of *C. retiniana, C. carnana-occidentalis,* and *C. lambertiana* elsewhere, blend in the Trinity-southern Cascades (De Benedictis, Part 4 this vol.).

Farther north, at the Oregon border, our 1966 collection from Mt. Ashland revealed that the situation is still more confused, with a continuous array of phenotypes, from *C. retiniana*-like to *C. carnana-* or *occidentalis*-like (see Figs. 7.51-7.56 in Powell and De Benedictis, Part 7 this vol.). In central Oregon, there are extremely variable populations; larvae range from green to brown, adults from somewhat *C. retiniana*-like, resembling those of Siskiyou Co., to large, broad-winged forms with females exhibiting dichromatism, gray and rust, genetic traits that are characteristic of the Spruce Budworm, *C. fumiferana*, in eastern North America and across northern regions to Alaska (Stehr 1954, 1967; UCB specimens). In eastern Oregon, however, populations referred to *C. occidentalis*, which we interpret as hybridizing, have intermediate forms rather than the clearcut rust and gray dimorphism known in *C. fumiferana*. A few individuals retain indications of the broad, pale banding of the forewing that is characteristic of *C. carnana* (Figs. 7.57-7.62 in Powell and De Benedictis, Part 7 this vol.).

Relationships West-to-East Across Southern Oregon

Research in the CANUSA-West program by W. J. A. Volney and his group at Berkeley was carried out concurrently with ours. They investigated populations along a transect across southern Oregon, from Mt. Ashland eastward to the Gearhart Mountain area near Bly. The forest composition is mixed *Pseudotsuga* and *Abies* at Mt. Ashland and Chiloquin Ridge east of Klamath Lake, while the eastern sites they sampled are outside the range of *Pseudotsuga*. There the forest resembles the Warner Mts. in tree species composition, with *Abies concolor* the only available Abietoid host.

In an elegant series of studies, Volney *et al.* (1983a, b, 1984), Liebhold et al. (1984) and Liebhold and Volney (1984) clearly demonstrated that the criteria which had been used to distinguish *Choristoneura retiniana* (=*viridis* Freeman) and *C. occidentalis*, namely forewing color and pattern (Freeman 1967) and larval colors (Harvey and Stehr 1967) do not represent reproductively isolated species in southern Oregon. They found that these characters are affected by temperature during larval feeding, as well as by genetic background, including sex-linked traits. In the western part of their transect, brown and green larval morphs and intermediates occurred on both *Abies* and *Pseudotsuga*, with the rank proportion of brown morphs directly correlated with the rank proportion of *Pseudotsuga* in mixed tree stands. Correspondingly, at the eastern sites, in absence of *Pseudotsuga*, all larvae were green. In laboratory matings, neither larval host nor larval color influenced mating success and fecundity.

These findings were supported by tests of pheromone specificity: using a method of scoring to characterize forewing pattern and color (Volney *et al.* 1983b), they found that 85% of males attracted by brown-larvae females and 82% of males attracted by green-larvae females were successfully classified as consistent with the correct female type. This indicates some field cross-attraction of adults formerly classified as *C. retiniana* and *C. occidentalis* in that region. Females reared from green, brown, and intermediate morph larvae attracted groups of males that were significantly different from each other in the percentage classified as "brown"-type (=*C.* "*occidentalis*"). Hybrid females produced from lab crosses of progeny from field-collected green and brown larvae, and F_2 backcrosses, attracted fewer males that were classified as "brown"-type, but not significantly different proportions, as compared with females reared from brown larvae (Liebhold *et al.* 1984). These findings support the conclusion that successful hybrid matings occur naturally in southern Oregon between populations that have been classified as *C. retiniana* and *C. occidentalis* (sensu Freeman 1967, Harvey and Stehr 1967).

Relationships South-to-North from Northern California to Central Oregon

In July-August 1982, we sampled populations during 1- to 8-night periods at a series of localities extending in an arc from the Klamath Ranges of northwestern California, northeastward to east-central Oregon (Fig. 2.3). We deployed synthetic pheromones and virgin females of *Choristoneura retiniana* from Tehachapi Mountain and of *C. occidentalis* from a lab colony maintained by the Pacific Southwest Forest Experiment Station, Berkeley.

In Siskiyou Co., California, the pheromones attracted recognizably different males, although the distinctness of forewing pattern and forewing breadth was less than in the Sierra Nevada (Table 2.8). At Mt. Ashland, traps baited with *C. retiniana* females and RET-C lure (n=7) attracted 58 males, while those with *C. occidentalis* females and bait (n=7) caught 124. Although we perceived differences between the two groups of males, these were not consistent, and many individuals are not classifiable. The prevailing color in both groups is brick-red or rust, as in *C. carnana californica*; none of the males is gray-brown, a characteristic of males in *C. occidentalis* to the north and of many males in typical *C. retiniana* in the Mt. Shasta-Modoc region. In this respect, this hybrid population at Mt. Ashland shows closer affinity to *C. carnana* than to *C. occidentalis*.

West and east of Klamath Lake (Lake of the Woods, Chiloquin Ridge), only 45-85 km northeast of Mt. Ashland, there are no consistent phenotypic differences between males attracted to pheromones of *Choristoneura retiniana* (9 traps, 101 males) and *C. occidentalis* (9 traps, 105 males). Moreover, 19 males of a similar phenotypic range were attracted to SUBRET-B bait (6 traps). Males at *C. occidentalis* females and bait were generally darker than at Mt. Ashland, mostly rust- to dark reddish brown, and mostly dark reddish brown to dark brown at *C. retiniana* females and bait. Northward, populations on the east slope of the Cascade Range (northwest of Sisters) and in the Aldrich-Strawberry Mts. (Canyon Creek Summit, Moco Springs) exhibit similar characteristics. An indistinguishable conglomeration of phenotypes was attracted to all three pheromone types. Males are gray-brown; reared females from these populations range from rust- (25%) to gray-brown (60%). In these areas, about 10 males per trap (n=10) of the same phenotypic range were attracted to SUBRET-B lure.

In total, traps averaged 10.3 males per night (n=93) in east-central and south-central Oregon, with females of *C. occidentalis* and *C. retiniana* and RET-C bait attracting about average numbers, OXY bait attracting significantly more ($\bar{x}$= 16.9), and SUBRET-B lure fewer ($\bar{x}$= 5.9) than the mean.

The data suggest that *Choristoneura lambertiana* may be involved in the hybridization; this question has not been addressed experimentally. Alternatively, males in the *C. retiniana-occidentalis* hybrid populations may include individuals that respond to a much broader array of 14- carbon aldehyde, acetate, and alcohol compounds than do males in non-hybridizing populations. Four SUBRET-B lure traps that we deployed in the Moco Springs-Delintment Lake area, Harney Co., Oregon, attracted 13 males of the hybrid phenotypes (11% of the 20-trap sample). However, specimens of *C. l. subretiniana* (n=31) showing no indication of hybrid characteristics were reared in 1945 in the same area, and one of the phenotype was taken next to its pupal exuvium in 1982 on *Pinus Ponderosa*. This suggests that *C. lambertiana* maintains its integrity in that region, and catches at SUBRET reflect inability of males in the hybridizing populations to recognize differences between the pheromones with similar principal components.

On the west side of the Cascade Range (4 sites in Linn and Douglas Cos.), blacklight (n=2) and pheromone traps (n=14 OXY + *C. occidentalis* females), which we deployed for single nights, attracted a relatively monomorphic, gray *C. occidentalis* (n=54). At these localities, no *Choristoneura* came to traps baited with RET-C (n=l0), female *C. retiniana* (n=4), or SUBRET lure (n=10). In marked contrast, hybridizing populations were encountered northwest of Sisters, Deschutes Co., east of the Cascade Range crest, only 27 airline km northeast of Alder Springs, Linn Co., where the monomorphic *Choristoneura* responded only to female *C. occidentalis* and OXY lure. At the Sisters locality, 9 traps (4 RET+ *C. retiniana* females, 3 OXY+ *C. occidentalis* females, 2 SUBRET) attracted 169 moths of a conglomeration of phenotypes to all traps.

Convergence in Phenotypes Between Males Attracted to Choristoneura retiniana and C. carnana Pheromones, from Southern California to Oregon

We randomly selected subsamples of l0 males attracted either to RET-C bait and *C. retiniana* females or to OXY bait and *C. carnana* and *C. occidentalis* females, at various localities in California and Oregon from l980 to l982. Each specimen was scored for five wing characters: forewing (FW) breadth (length:width ratio); FW banding; FW black flecks; % dimorphism in FW color (tan vs. mottled); and hindwing (HW) color. Each character was ranked from l.0 (most *C. carnana carnana*-like) to 4.0 (most like *C. retiniana* in southern California), and a value was assigned to each specimen based on comparison to standard exemplars. The five resultant scores were totalled to provide a phenetic index of the specimen (i.e., a possible range of 5.0 for *C. carnana carnana* to 20.0 for *C. retiniana* with l00% tan morph). A mean value was then calculated for each population sample. These scores have no statistical interpretation; they merely illustrate degree of likeness that sympatric populations possess when compared to one another.

The results are depicted in Figure 2.6. There is no *a priori* judgment of species; the triangles represent mean values for males attracted to OXY bait or *Choristoneura carnana* or *occidentalis* females, and the circles represent those that came to RET-C bait or *C. retiniana* females.

There are obvious differences in phenotypes in southern California and in the Sierra Nevada (compare Figs. 7.29-7.32 and 7.33-7.40 in Powell and De Benedictis, Part 7 this vol.), but northward, sympatric population samples and latitudinally comparable allopatric samples converge in phenetic scores. There are recognizable differences in southern Siskiyou Co. (Trail Cr., Hockaday Spr., Mt. Shasta), but scarcely so at Prather Creek, 50 km north of Mt. Shasta, and even less so at the Oregon border (Mt. Ashland) (ibid., Figs. 7.51-7.56). These are mean values, so there is broad overlap in the phenotypic expression of the characters among individuals, such that in eastern Oregon, from Klamath Lake northward, there is no discernible difference in random samples nor in pheromone-based collections of males or groups segregated when adults are reared from green, brown, or intermediate larvae.

SUMMARY AND CONCLUSIONS

Species of Conifer-feeding *Choristoneura* in the Western United States

This work demonstrates that there are two complexes of Abietoideae-feeding *Choristoneura* that are sympatric in widespread areas of California, Nevada and Utah. A third complex, which is primarily restricted in larval feeding to Pinoideae, is sympatric with one or two members of the Abietoideae series in some populations, but it was not emphasized in this study.

Choristoneura retiniana (Walsingham) feeds on *Abies*, has unpigmented larval integument (appearing green in life), weakly pigmented pupal integument, and dimorphic color phenotypes in both males and females. The principal component of its sex pheromone is E-11 tetradecenyl acetate. This species occurs in the mountains of California, Great Basin ranges, and Wasatch Range of north-central Utah. Populations of the Sierra Nevada, Tehachapi, Transverse, and Peninsular Ranges in California and from the Great Basin Ranges in Nevada and Utah have consistently narrow forewings and are dimorphic in both sexes, but relatively uniform in adult phenotype, and the name *spaldingiana* Obraztsov can be applied to them in a subspecific context.

Choristoneura carnana (Barnes and Busck) feeds on *Pseudotsuga*, has heavily pigmented larval integument (appearing mottled, brick-red in life), heavily pigmented pupal integument, and monomorphic, predominantly rust colored adults. Populations in southern California (*carnana carnana*) feed on *P. macrocarpa*, those elsewhere feed on *P. menziesii*. The principal component of the sex pheromone is E-11-tetradecenal. *C. carnana* occurs in the Transverse Ranges, Sierra Nevada, Santa Cruz Mts., North Coast, Klamath, and southern Cascade Ranges of California.

Choristoneura occidentalis Freeman feeds on various Abietoideae (Stehr 1967), and in the northwestern states has more variable pigmentation of larvae, pupae and adults (Harvey and Stehr 1967, Freeman 1967) than the two preceding species show in California. In the areas encompassed by this study, *C. occidentalis* was associated with *Abies lasiocarpa* in northern Nevada, with mixed stands of *Abies* and *Pseudotsuga* in eastern Nevada and northern Arizona, and with mixed *Abies-Picea-Pseudotsuga* forests in Washington, Oregon, and Utah. The principal component of the pheromone is E-11-tetradecenal; evidently it is very similar to that of *C. carnana*.

Choristoneura lambertiana (Busck) feeds on *Pinus*, has heavily pigmented larval and pupal integument, and is relatively monomorphic in adult color within any given population. The adult phenotype is geographically polytypic, and the previously named entities outlined below are regarded as allopatric subspecies, on the assumption that all the pine-feeding populations in the western United States represent one species.

C. l. lambertiana is known only from south-central Oregon and adjacent Siskiyou Co., California, where it has been detected feeding on *Pinus lambertiana* at two localities.

C. l. subretiniana Obraztsov occurs in the Transverse Ranges of southern California, along the crest and east side of the Sierra Nevada, and in the Warner Mts. of northeastern

California, associated with its primary host, *Pinus contorta murrayana*. Populations associated with *P. ponderosa* in eastern Oregon differ in forewing pattern but have been assigned to this subspecies.

C. l. ponderosana Obraztsov occurs on the east slope of the Rocky Mts. in Colorado, in New Mexico, and in the Black Hills of eastern Wyoming-western South Dakota (Obraztsov 1962, Stevens *et al.* 1977). There are similar forms in the Basin Ranges of southern Utah, northern Arizona, and southern Nevada. Populations in western Wyoming, Montana, and northward, which display varying phenotypes typical of neither *ponderosana* nor *subretiniana*, have been referred to *C. l. ponderosana* (Stevens *et al.* 1977) or *C. l. lambertiana* (Obraztsov 1962, Stehr 1967), or have been unassigned to subspecies (Harvey 1985; De Benedictis, Part 4 this vol.). Recorded larval associations are with *Pinus ponderosa* in Colorado, South Dakota, and Wyoming, and with *P. flexilis* and *P. contorta murrayana* in Montana.

Males of *C. l. ponderosana* and *C. l. subretiniana* both respond to a synthetic pheromone with the principal components a 90:10 blend of E:Z-11 tetradecenyl acetate and E:Z-11 tetradecenyl alcohol (Daterman *et al.*, Part 3 this vol.).

Coastal populations in Washington and Oregon reared from or associated with *Pinus contorta contorta* are dissimilar by facies and uncus shape to any named *Choristoneura lambertiana* subspecies (De Benedictis, Part 4 this vol.). Their synthetic pheromone attraction has not been tested, and their taxonomic affinity is uncertain. Gray and Slessor (1989) reported a population at Richmond, British Columbia, near Vancouver, where larvae fed on Scots Pine, *P. sylvestris*, an introduced Palaearctic tree. They speculated that the *Choristoneura* had originated via introduction of Scots Pine to the district; it seems likely to us that the population is conspecific with coastal populations associated with *P. contorta* to the south. The British Columbia population produces a pheromone that contains an unique blend of E:Z-11 acetates and alcohols and therefore Gray and Slessor interpreted the population as a reproductively isolated, undescribed species. However, their bioassay data contradict the assumption because males were attracted in appreciable numbers by unmated females of *C. pinus* from an unstated source population, as well as by lures of various blends of the principal components, which are similar to those of *C. lambertiana* (Daterman *et al.*, Part 3 this vol.).

The coastal populations of the Pacific Northwest may be a western race of *C. pinus*.

Sympatric Interrelationships

Data presented above and compiled from other specimen sources indicate the following associations of sympatry in the region examined during this study:

Choristoneura retiniana and *C. carnana*: San Gabriel and San Bernardino Mts., west slope of the Sierra Nevada below ca. 1,800 m, east and west sides of northern Sierra Nevada and southern Cascade Range in Shasta and Siskiyou Cos., California.

C. retiniana and *C. occidentalis*: Schell Creek and Snake Ranges, White Pine Co., Nevada; Wasatch Range and mountains of southern Utah.

C.retiniana and *C. lambertiana subretiniana:* Warner and Manzanita Mts., Modoc Co., northern Sierra Nevada at 1,600 m and central Sierra Nevada above ca. 1,900 m.

C. retiniana and *C. l. ponderosana*: Charleston Mt., Clark Co., Nevada; southern Utah.

C. occidentalis and *C. l. ponderosana*: Colorado, New Mexico, southern Utah and northern Arizona.

Three species, two Abietoideae-feeding and one Pinoideae-feeding, occur together in a narrow zone east of the northern Sierra Nevada and in southern Utah.

In the western United States south of 42° N lat., *Choristoneura retiniana* and either *C. carnana* or *C. occidentalis* are sympatric only in mixed *Abies-Pseudotsuga* forests. Our data indicate that host tree specialization aids in maintenance of species separation, *C. retiniana* feeding on *Abies*, *C. carnana-occidentalis* primarily on *Pseudotsuga*. In areas where only *Abies* occurs, either but not both *Choristoneura* species is present (e.g., *C. retiniana* in Modoc Co., Tehachapi Mts., and Charleston Mt.; *C. occidentalis* in Copper-Jarbidge Mts.). *Pseudotsuga* growing in absence of *Abies* supports the *C. carnana-occidentalis* complex (e.g., California Coast Ranges) but not *C. retiniana*. Either *C. carnana* or *C. occidentalis* may be present in mixed Abietoideae stands (*C. carnana* in Klamath Ranges, Trinity Co.; *C. occidentalis* in eastern Nevada, northern Arizona), but we did not discover any *Abies-Pseudotsuga* site exclusively occupied by *C. retiniana*..

Thus, available evidence indicates that where a member of the *C. carnana-occidentalis* complex is sympatric with *C. retiniana*, each is host specific. In places where *C. occidentalis* is released from competition with *C. retiniana*, it is polyphagous on various genera of Abietoideae. The same relationship might be shown by *C. carnana*, but in most of its range it is restricted to lower elevations where only *Pseudotsuga* occurs. In the one area where we found *C. carnana* in mixed *Abies-Pseudotsuga* forest in absence of *C. retiniana*, northern Trinity Co., both hosts were used.

Reproductive Isolation

Sex pheromone chemistry appears to be the primary reproductive isolation mechanism in sympatric populations. Interspecific matings resulting in fertile F_1 families may be obtained in the laboratory among various *Choristoneura* (Volney, Liebhold, pers. correspondence), but none of the sympatric population pairs discussed here have been tested rigorously. Interpopulational attraction tests with virgin females showed 100%, and synthetic pheromone tests in sympatry showed a >99%, separation of males attracted in the field. There is no evidence from our larval collections or from adult specimens that hybridization occurs in natural populations in the areas south of 42° N. lat. that were examined in this study.

Northward, however, starting at about the Oregon and Idaho border (42° N lat.), the reproductive isolation breaks down. The distinct adult and larval morphs of Californian *Choristoneura retiniana* and *C. carnana* disappear and are replaced by heterogeneous populations having green to brown larvae and variable forewing patterns. Populations of this nature, which occur throughout the mountain ranges of the Columbia Plateau of eastern Oregon, lack the tan phenotype of *C. retiniana*, but exhibit pronounced sexual dimorphism, with a range in colors from gray to rust-brown in females, as is also characteristic of *C. fumiferana.* In the areas east of the Cascade Range in central Oregon, the three synthetic pheromones that males of *C. retinana, C. carnana*, and *C. lambertiana* distinguish in California, Nevada, and Utah, all attract a similar phenotypic range of males. We do not know if individual males react to a broad spectrum of E-11-tetradecenyl aldehyde and E-11-tetradecenyl acetates, so that the conglomeration that appears at any given female or synthetic pheromone is a random sample of males in the vicinity, or whether each male is genetically predisposed to respond to one kind of pheromone, and some degree of assortative mating occurs. Preliminary tests with hybrid females suggest the latter, with aldehyde dominance (De Benedictis *et al.*, Part 5 this vol.). Discriminant analysis of phenotypes among attracted males in southern Oregon (Volney *et al.* 1984, Liebhold *et al.* 1984) demonstrate that consistent arrays of males are attracted to females of known larval colors, indicating some differential attraction. This suggests a preference by individual males to one pheromone blend over another. A portion of the population may be indiscriminate, with a larger portion discriminating.

Similar tests of comparative attraction have not been carried out in Idaho and eastern Washington, but the degree of variation observed in specimens (e.g., Harvey and Stehr 1967) suggests that some hybrid mixing occurs throughout the intermountain region. Unfortunately from a nomenclatural standpoint, the type locality of *Choristoneura occidentalis* is in the Simcoe Mts. in Klickitat Co., in south-central Washington, and thus may be within the hybrid zone. Nonetheless, phenotypic variation in larvae and adults in populations of the east side of the Cascade Range in Washington is not as heterogeneous as in eastern Oregon. In 1988 we obtained reared adults from Satus Pass, Klititat Co. (n=4) and from Okanogan Co. (n=137). Females attracted *C. carnana* and not *C. retiniana* in the Sierra Nevada, but only one (of three deployed) was from Klititat Co. (De Benedictis *et al.*, Part 5 this vol.).

Populations on the west side of the Cascades appear to be less variable, and males respond to OXY and not RET lure, evidently also traits of the Rocky Mts. populations.

Conventional use in the literature has referred all of the budworm populations associated with Abietoideae in the northwestern United States and Rocky Mts. to *Choristoneura occidentalis* (sensu Freeman 1967). Clearly, however, the hybridizing populations of eastern Oregon should be excluded from this designation. Probably this is also true for several populations in the Great Basin ranges (e.g., Jarbidge Mts. and Schell Creek Range, Nevada) and the Flagstaff area of northern Arizona, where three Abietoideae-feeding entities, none of which is *C. retiniana,* evidently coexist.

ACKNOWLEDGMENTS

Cooperation by G. E. Daterman and his associates, U.S. Forest Service, Corvallis, Oregon, in providing pheromone baits made this study possible. Nils Westerland, Tahoe City, California, carried out blacklight trapping during 1974-1981, providing flight period and relative abundance data for *Choristoneura retiniana* and *C. lambertiana subretiniana*. He and Florence Westerland cooperated by monitoring pheromone traps and providing us with hospitality during the 1979-1981 field seasons. R.C. Heald and other authorities in the former Department of Forestry, University of California, Berkeley, arranged for us to use facilities at the Blodgett Forest Research Station.

LITERATURE CITED

Campbell, I.M. 1967. On coniferophagous species of *Choristoneura* (Lepidoptera: Tortricidae) in North America. IV. Sexual isolation between three species. Canad. Entomol. 99:482-486.

Daterman, G.E., H.T. Cory, G.D. Daves, Jr., R.E. Stevens, and D.G. Fellin. 1983. A preliminary report on the sex pheromones of three pine-feeding budworms of the *Choristoneura lambertiana* Busck complex in the western United States; 17 pp. On file: Forestry Sciences Laboratory, Corvallis, Oregon.

Daterman, G.E., H.T. Cory, L.L. Sower, and G.D. Daves, Jr. 1984. Sex pheromone of a conifer-feeding budworm, *Choristoneura retiniana* Walsingham. J. Chem. Ecol. 10:153-160.

Daterman, G.E., L.L. Sower, R.E. Stevens and D.G. Fellin. 1995. Pheromone chemistry and response behavior of conifer-feeding *Choristoneura* in the western United States. Part 3, this volume.

De Benedictis, J.A. 1995. Phenetic studies of spruce budworm populations (*Choristoneura* species). Part 4, this volume.

De Benedictis, J. A., A. M. Liebhold and J. A. Powell 1995. Studies of pheromone attraction and inheritance using virgin females of *Choristoneura* species, in the central Sierra Nevada, California. Part 5, this volume..

Freeman, T.N. 1958. The Archipinae of North America (Lepidoptera: Tortricidae). Canad. Entomol. 90 (Suppl.7); 89 pp.

Freeman, T.N. 1967. On coniferophagous species of *Choristoneura* (Lepidoptera: Tortricidae) in North America. I. Some new forms of *Choristoneura* allied to *C. fumiferana*. Canad. Entomol. 99:449-455.

Gray, T. G., and K. N. Slessor. 1989. Morphology, life history and identification of sex pheromone components of an undescribed species of *Choristoneura* (Lepidoptera: Tortricidae) on Scots Pine in British Columbia. J. Entomol. Soc. Brit. Col. 86:39-47.

Harvey, G. T. 1985. The taxonomy of the coniferophagous *Choristoneura* (Lepidoptera: Tortricidae): A review. Pp. 16-48 *in* C. J. Sanders, *et al.* (eds.), Recent advances in spruce budworms research. Proc. CANUSA Spruce Budworms Res. Symp., Bangor, Maine, 1984.Canad. For. Serv.; Ottawa; 527 pp.

Harvey, G. T., and G. Stehr. 1967. On coniferophagous species of *Choristoneura* (Lepidoptera: Tortricidae) in North America. III. Some characters of immature forms helpful in the identification of species. Canad. Entomol. 99:464-481.

Liebhold, A.M., W.J.A. Volney, and W.E. Waters. 1984. Evaluation of cross-attraction between sympatric *Choristoneura occidentalis* and *C. retiniana* (Lepidoptera: Tortricidae) populations in south-central Oregon. Canad. Entomol. 116:827-840.

Liebhold, A.M., and W.J.A. Volney. 1984. Effect of temporal factors on reproductive isolation between *Choristoneura occidentalis* and *C. retiniana* (Lepidoptera: Tortricidae). Canad. Entomol. 116:991-1005.

Lyon, R.L., C.E. Richmond, J.L. Robertson, and B.A. Lucas. 1972. Rearing diapause and diapause-free western spruce budworm (*Choristoneura occidentalis*) (Lepidoptera: Tortricidae) on an artificial diet. Canad. Entomol. 104:417-426.

Obraztsov, N.S. 1962. New species and subspecies of North American Archipini, with notes on other species (Lepidoptera: Tortricidae). Amer. Mus. Novitates 2101:1-26.

Powell, J.A. 1964. Biological and taxonomic studies on tortricine moths, with reference to the species in California (Lepidoptera: Tortricidae). Univ. Calif. Publ. Entomol. 32:1-317.

Powell, J.A. 1980. Nomenclature of Nearctic conifer-feeding *Choristoneura* (Lepidoptera: Tortricidae): Historical review and present status. Gen. Tech. Rep. PNW-100, USDA For. Serv.:1-18.

Powell, J.A. 1983. Tortricidae. Pp. 31-42 in Hodges, R.W. (ed.) Check List of the Lepidoptera of America North of Mexico. E. W. Classey and Wedge Entomol. Found., London.

Powell, J.A. 1995. Introduction: Historical review, taxonomic problems, research approaches. Part 1, this volume.

Powell, J.A., and J.A. De Benedictis. 1982. Taxonomic relationships and pheromone isolation among western spruce budworm populations. Canada/U.S. Spruce Budworms Program-West Final Report; 41 pp. [unpublished].

Powell, J.A., and J.A. De Benedictis. 1995. Evolutionary interpretation, taxonomy, and nomenclature. Part 7, this volume.

Sanders, C.J. 1971. Daily activity patterns and sex pheromone specificity as sexual isolating mechanisms in two species of *Choristoneura* (Lepidoptera: Tortricidae). Canad. Entomol. 103:498-502.

Smith, S.G. 1953. Reproductive isolation and the integrity of two sympatric species of *Choristoneura* (Lepidoptera: Tortricidae). Canad. Entomol. 85:141-151.

Stark, R.W., and J.H. Borden. 1965. Life history of *Choristoneura lambertiana subretiniana* Obraztsov (Lepidoptera: Tortricidae) attacking lodgepole pine. Canad. Entomol. 97:684-690.

Stehr, G. 1954. A laboratory method for rearing the spruce budworm, *Choristoneura fumiferana* (Clem.) (Lepidoptera: Tortricidae). Canad. Entomol. 86:423-428.

Stehr, G. 1967. On coniferophagous species of *Choristoneura* (Lepidoptera: Tortricidae) in North America. II. Geographic distribution in accordance with forest regions. Canad. Entomol. 99:456-463.

Stevens, R.E., T.K. Borg, and T.O. Thatcher. 1977. Notes on a pine-feeding budworm, *Choristoneura lambertiana ponderosana* (Lepidoptera: Tortricidae), in the Colorado Rockies. Canad. Entomol. 109:1269-1274.

Volney, W.J.A., W.E. Waters, R.P. Akers, and A.M. Liebhold. 1983a. Variation in spring emergence patterns among western *Choristoneura* spp. (Lepidoptera: Tortricidae) populations in southern Oregon. Canad. Entomol. 115:199-209.

Volney, W.J.A., A.M. Liebhold, and W.E. Waters. 1983b. Effects of temperature, sex, and genetic background on coloration of *Choristoneura* spp. (Lepidoptera: Tortricidae) populations in south-central Oregon. Canad. Entomol. 115:1583-1596.

Volney, W.J.A., A.M. Liebhold, and W.E. Waters. 1984. Host associations, phenotypic variation, and mating compatibility of *Choristoneura occidentalis* and *C. retiniana* (Lepidoptera: Tortricidae) populations in south-central Oregon. Canad. Entomol. 116:813-826.

Weatherston, J., W. Roelofs, A. Comeau, and C.J. Sanders. 1971. Studies of physiologically active arthropod secretions. 10. Sex pheromone of the eastern spruce budworm, *Choristoneura fumiferana* (Lepidoptera: Tortricidae). Canad. Entomol. 103:1741-1747.

Table 2.1. Test of attractiveness by females of *Choristoneura carnana californica* following refrigeration (larvae collected 21 May 1980, traps deployed 17-20 June 1980, 1 km SE and 3 km NE Angwin, Napa Co., California)

Female #	Days pupa refrigerated	Days moth refrigerated	Males attracted
1	--	14	2
2	--	11	4
3	--	8	24
4	--	7	24
5	--	7	15
6	--	6	25
7	7	6	18
8	7	5	12
9	10	2	3
10	10	0	20

Table 2.2. Source and test site data for cross-population attractiveness experiments with virgin female *Choristoneura*

Species	Larval Source	Deployed at	Trap Numbers
1980			
carnana carnana	San Gabriel Mts.	Blodgett	3
carnana californica	Lake Co.	Blodgett	13
"	Mendocino Co.	Blodgett	2
"	Trinity-Siskiyou Co.	Blodgett	9
"	"	Modoc Co.	2
retiniana	Modoc Co.	Blodgett	1
"	"	Schell Cr. Range	1
"	"	Snake Range	1
retiniana	McCloud Summit, Siskiyou Co.	Blodgett	1
retiniana	Tehachapi Mt.	Blodgett	12
"	"	Modoc Co.	7
1981			
retiniana	Tehachapi Mt.	Bucks Lk., Plumas County	2
"	"	Sulphur Cr., Plumas County	1
"	"	Tahoe City	4

Table 2.3. Designations and chemical formulations of synthetic pheromone baits used in 1978-1982 to attract males of conifer-feeding *Choristoneura*

Lure	Species Attracted	Dates Deployed	Chemical Formulation	Results
"OXY"	*carnana* *occidentalis*	1978-82	92:8 E:Z-11-tetradecenyl aldehyde	effective
"RET-A"	none	1978	100% E-11-tetradecenyl acetate	not effective
"	none	1979	95% E-11-tetradecenyl acetate, 5% E-11-tetradecenyl alcohol	not effective
"RET-B"	*retiniana*	1980	95% E-11-tetradecenyl acetate, 5% Z-11-tetradecenyl acetate	weakly effective
"RET-C"	*retiniana*	1981	94% 93:7 E:Z-11-tetradecenyl acetate, 6% 92:8 E:Z-11-tetradecenyl alcohol	effective
"	"	1982	90% 92:8 E:Z-11-tetradecenyl acetate, 10% 90:10 E:Z-11-tetradecenyl alcohol	effective
"SUBRET-A"	none	1980	70% Z-11-tetradecenyl acetate, 20% E-11-tetradecenyl acetate, 10% 70:20 Z:E-11-tetradecenyl alcohol	not effective
"SUBRET-B"	*lambertiana*	1981	90% 60:40 E:Z-11-tetradecenyl acetate, 10% 67:33 E:-11-tetradecenyl alcohol	effective
"	"	1982	90% 60:40 E:Z-11-tetradecenyl acetate, 10% 60:40 E:Z-11-tetradecenyl alcohol	effective

Table 2.4. Summary of hostplant associations of *Choristoneura* in California and Oregon, based on adults reared from field-collected larvae, during 1970-1982 (see Appendix 2.1 for data)

Species	Hostplant	No. of collections
C. carnana carnana	*Pseudotsuga macrocarpa*	4
C. carnana "Sierra Nevada"	*Pseudotsuga menziesii*	4
	Abies concolor	1
C. carnana californica	*Pseudotsuga menziesii*	20
	Abies concolor	4
C. retiniana retiniana	*Abies concolor*	42
C. retiniana spaldingiana	"	9
	Abies magnifica	3
C. hybrid populations (Oregon)	*Abies concolor*	5
	Abies magnifica	1
C. lambertiana subretiniana	*Pinus contorta murrayana*	2
	Pinus albicaulis (?)	1
	Abies concolor	4
	Abies magnifica	1

Table 2.5. *Choristoneura* males taken in traps baited with synthetic pheromones and virgin females during 5 seasons in the Blodgett Forest area, El Dorado Co., California

Bait	No. of male moths trapped (No. traps)	retiniana	carnana	Avg./trap
1978:				
"RET-A"	(18)	0	0	-
"OXY"	(18)	0	70	3.9
1979:				
"RET-A"	(11)	0	0	-
"OXY"	(12)	0	201	16.7
1980:				
"RET-B"	(6)	19	0	6.3
♀ *retiniana*	(12)	62	0	5.2
Tehachapi Mts.				
"OXY"	(6)	0	66	11.0
♀ *carnana carnana*	(3)	0	21	7.0
San Gabriel Mts.				
♀ *c. californica*	(15)	0	65	4.3
Lake-Mendocino Cty.				
♀ *c. californica*	(9)	0	122	13.5
Trinity-Siskiyou Cty.				
1981:				
"RET-C"	(21)	325	0	15.5
"OXY"	(20)	0	101	5.0
1982:				
"RET-C"	(12)	66	0	5.5
"OXY"	(12)	0	71	5.9
Totals		472	718	

Table 2.6. *Choristoneura* males taken in traps baited with synthetic pheromones and virgin females during 4- to 6-week periods, July-August, 1981, and 6- to 12-week periods, May-August, 1982, in Plumas and Placer counties, California

		No. of males moths attracted			
Bait	(No. traps)	*retiniana*	*camana*	*l. subretiniana*	Avg./trap
Bucks Lake (1981):					
"RET-C"	(2)	46	0	0	23
♀ *retiniana*	(2)	2	0	0	1
Tehachapi					
"OXY"	(2)	0	0	0	
"SUBRET-B"	(2)	0	0	61	30.5
Meadow Valley (1982):					
"RET-C"	(3)	21	0	0	7.0
"OXY"	(3)	0	21	0	7.0
"SUBRET-B"	(3)	0	0	0	
Big Creek Rd. (1981):					
"RET-C"	(3)	20	0	0	6.3
"OXY"	(3)	0	57	0	19
"SUBRET-B"	(3)	1	0	0	0.3
Sulphur Creek (1981-82):					
"RET-C"	(7)	63	0	0	9
♀ *retiniana*	(1)	9	0	0	9
Tehachapi					
"OXY"	(8)	0	122	0	15.2
"SUBRET-B"	(8)	0	0	2	0.25
Gold Lake Rd. (1981-82):					
"RET-C"	(5)	53	0	0	10.6
"OXY"	(5)	0	52	0	10.4
"SUBRET-B"	(5)	0	0	0	
Lake Tahoe (1981):					
"RET-C"	(2)	11	0	0	5.5
♀ *retiniana*	(4)	8	0	0	2.0
Tehachapi					
"SUBRET-B"	(4)	0	0	9	2.25
Totals		234	252	72	

Table 2.7. *Choristoneura* males taken in traps baited with synthetic pheromones or virgin females during 1- to 12-night tests in Nevada and Utah, in August 1980 and July 1981

Site & Bait (nights)	(No. traps)	No. of male moths attracted			
		retiniana	*occidentalis*	*l. ponderosana*	Avg./trap
Timber Cr., Nev. (12):					
"RET-B"	(2)	0	1	0	0.5
♀ *retiniana*	(1)	10	0	0	10
Modoc Co.					
"OXY"	(2)	0	98	0	49
Wheeler Peak Rd., Nev. (11):					
"RET-B"	(2)	0	0	0	
♀ *retiniana*	(1)	14	0	0	14
Modoc Co.					
"OXY"	(2)	0	80	0	40
Charleston Mt., Nev. (1):					
"RET-C"	(3)	57	0	0	19
"OXY"	(3)	0	0	0	
"SUBRET-B"	(3)	0	0	21	7
Lehman Cr., Nev. (1):					
"RET-C"	(3)	24	0	0	8
"OXY"	(3)	0	1	0	0.3
"SUBRET-B"	(3)	0	0	0	
Willis Cr.-Duck Cr., Utah (1):					
"RET-C"	(4)	42	0	0	10.5
"OXY"	(4)	0	7	0	1.75
"SUBRET-B"	(4)	0	0	17	4.25
Mt. Timpanogos, Utah (1):					
"RET-C"	(5)	33	0	0	6.6
"OXY"	(5)	0	23	0	4.6
Totals		180	210	38	

Table 2.8. *Choristoneura* males taken in traps baited with virgin females or synthetic pheromones in Siskiyou Co., California, during July and August 1982. Source populations of females: *C. retiniana*, Tehàchapi; *C. occidentalis*, lab colony originally from Idaho and Montana

		No. of Male Moths Attracted		Hybrid population
Site and Bait (nights)	(No. Traps)	*retiniana*	*carnana*	"occidentalis"
Trail Creek Campgr. (1):				
"RET-C"	(3)	7	0	-
♀ *retiniana*	(1)	0	0	-
"OXY"	(3)	0	45	-
♀ *occidentalis*	(2)	0	11	-
"SUBRET-B"	(3)	0	0	-
Hockaday springs (1):				
"RET-C"	(3)	23	0	-
♀ *retiniana*	(2)	3	0	-
"OXY"	(3)	0	33	-
♀ *occidentalis*	(2)	0	0	-
McBride Spr., Mt. Shasta, 1,400 m (8):				
"RET-C"	(3)	6	0	-
♀ *retiniana*	(2)	33	0	-
"OXY"	(3)	0	1	-
♀ *occidentalis*	(2)	0	1	-
Everitt Memorial, Mt. Shasta, 1,800 m (8):				
"RET-C"	(3)	42+	0	-
♀ *retiniana*	(1)	25	0	-
"OXY"	(3)	0	1	-
♀ *occidentalis*	(1)	0	1	-
Prather Cr. (7):				
"RET-C"	(5)	9	0	0
♀ *retiniana*	(2)	23+	0	0
"OXY"	(6)	0	0	57+
♀ *occidentalis*	(2)	0	0	25+
"SUBRET-B"	(3)	0	0	0
Totals		171+	93	82+

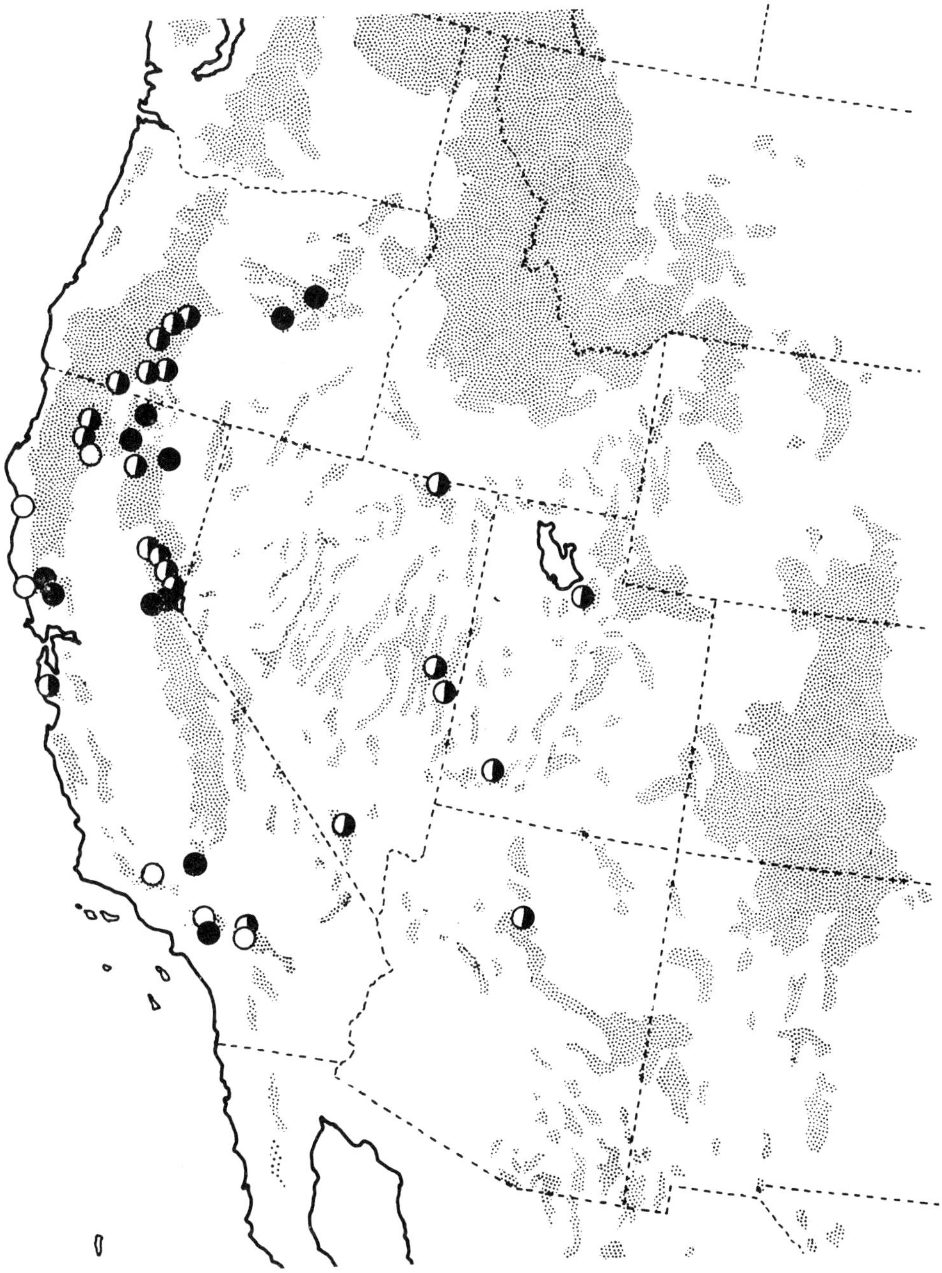

Fig. 2.1. Localities surveyed during 1978-1981, by larval collections, blacklight, and pheromone trapping. Open circles= larval collections; half-closed circles= pheromone and blacklight trap samples; closed circles= larval and pheromone/blacklight collections.

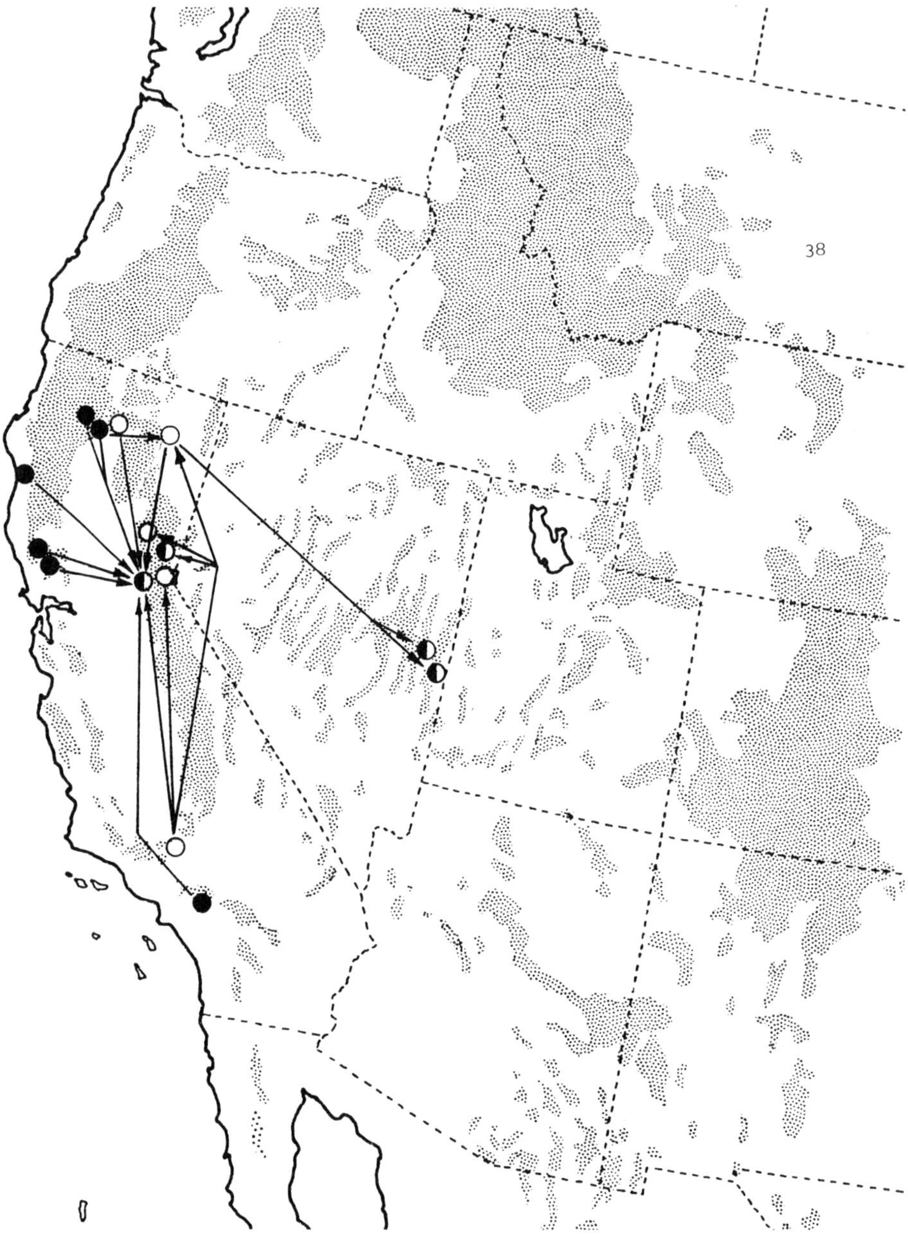

Fig. 2.2. Localities sampled during cross-population tests for attractiveness of virgin females to males in natural populations. Arrows indicate direction of transport of females. Open circles = *Choristoneura retiniana*; closed circles = *C. carnana*; half-closed circles = sympatric associations of two species at test site, *C. retiniana* and either *C. carnana* (California) or *C. occidentalis* (Nevada).

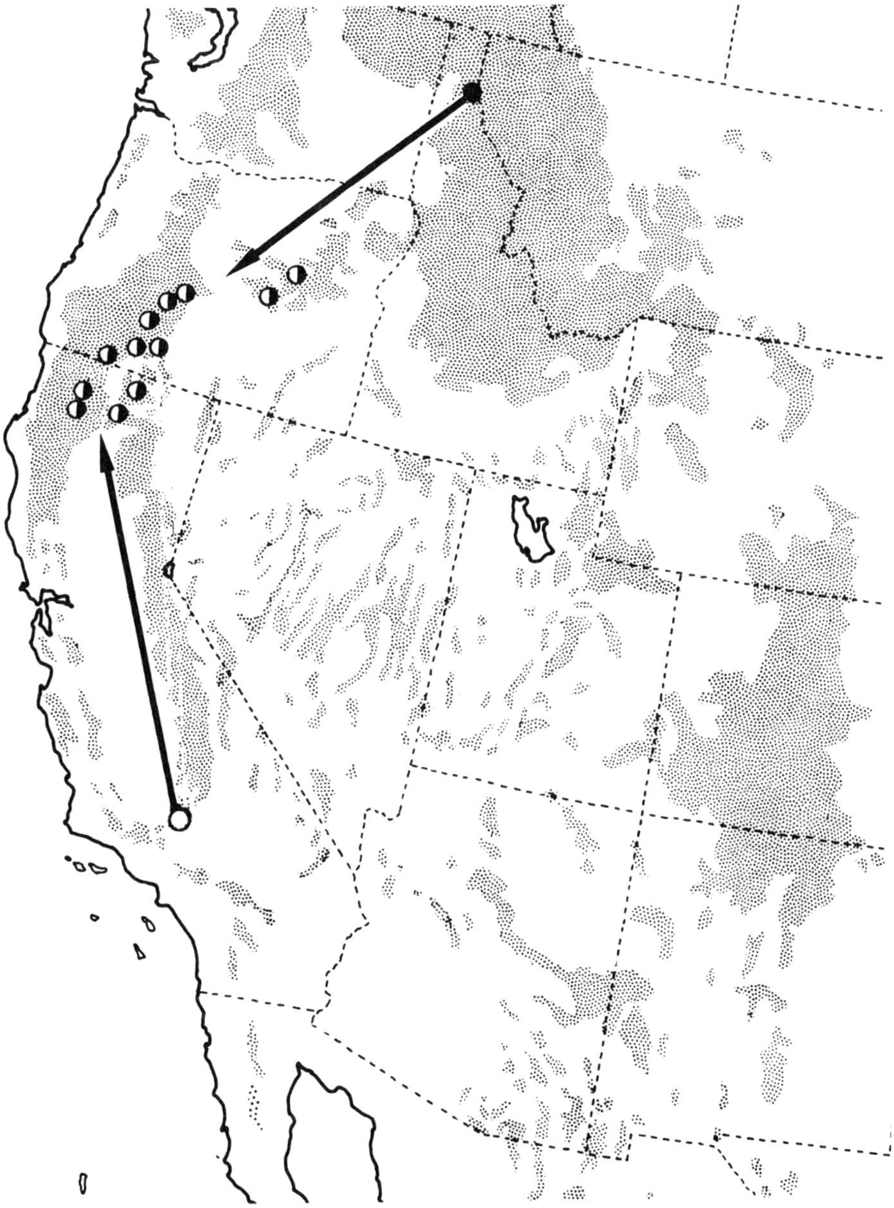

Fig. 2.3. Trap sample localities in an arc through the hybridization zone, from the Salmon Mts. in western Siskiyou Co., California., to the Aldrich-Strawberry Mts. near John Day, Grant Co., Oregon. We deployed virgin females of *Choristoneura retiniana* from Tehachapi Mtn. (open circle) and of *C. occidentalis* from northern Idaho-Montana (USFS lab colony at Berkeley) (closed circle), along with OXY and RET-C synthetic lures, for 1- to 8- day periods.

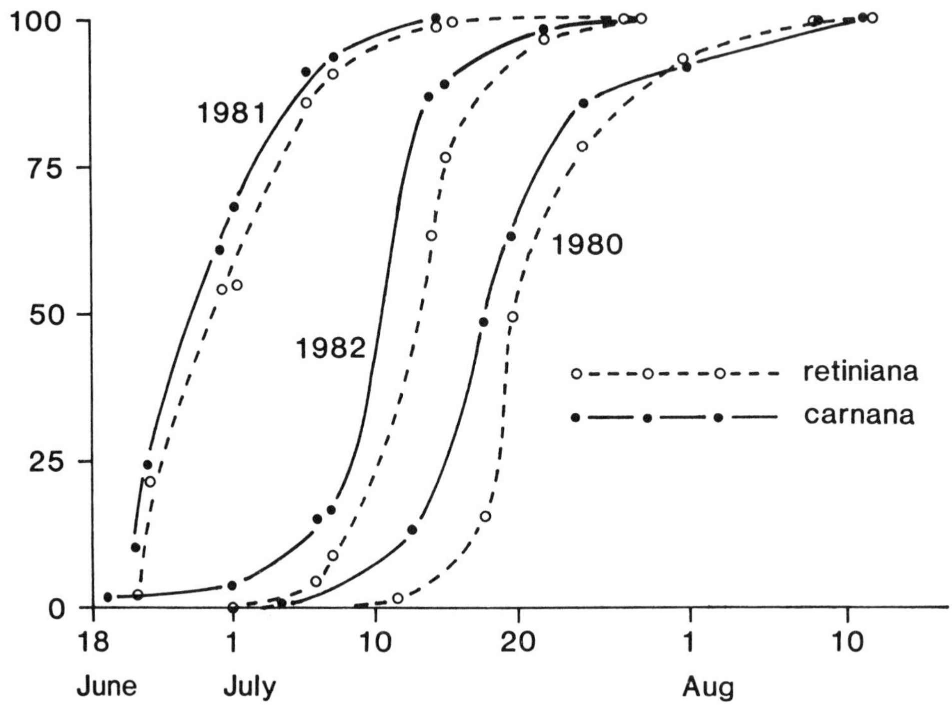

Fig. 2.4. Cumulative percentage of activity of *Choristoneura retiniana* and *C. carnana* during 3 seasons at Blodgett Forest, El Dorado Co., California, as indicated by continuous pheromone trap and weekly blacklight trap samples. In both years the mean of total catch for *C. carnana* preceded that of *C. retiniana* by 2-3 days, but the apparent lag in *C. retiniana* emergence in 1980 is believed to be partly artifactual because virgin females of *C. retiniana* were not available until 14 days after sampling began. The 1981 data do not reflect a sigmoid curve because flight had commenced prior to initiation of sampling.

C. retiniana: n = 140 (1980), 355 (1981), 78 (1982); *C. carnana*: n = 338 (1980), 130 (1981), 94 (1982).

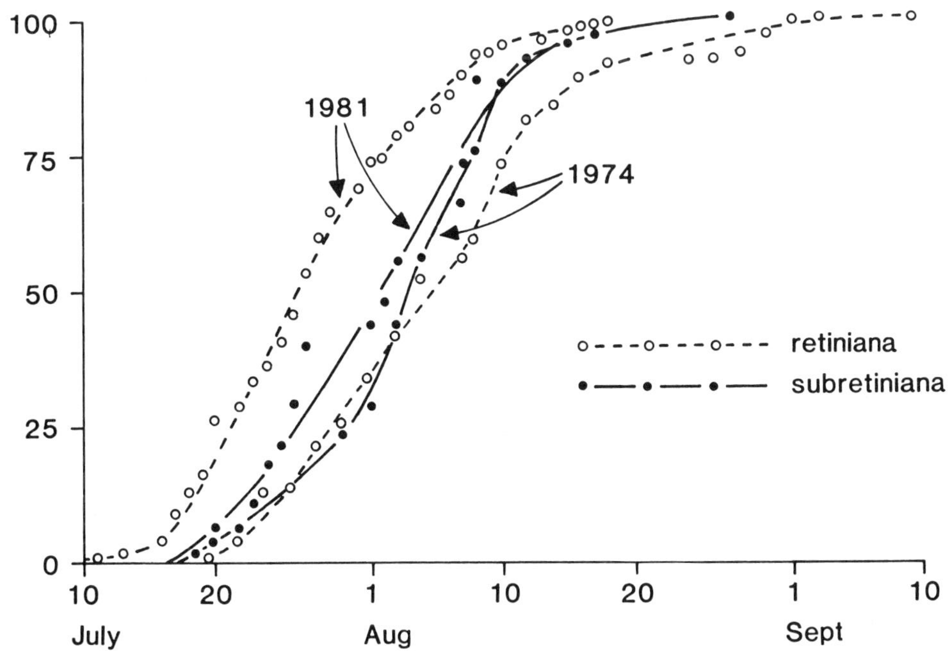

Fig. 2.5. Cumulative percentage of activity of *Choristoneura retiniana* and *C. lambertiana subretiniana* at Ward Creek, 3 mi S Tahoe City, Placer Co., California, as indicated by nightly blacklight trap sampling. The mean in total numbers in 1974 was reached for both species on August 5; while in 1981 *C. l. subretiniana* lagged about 6 days behind *C. retiniana* at the accumulation midpoint.

C. retiniana: n = 355 (1974), 301 (1981); *C. l. subretiniana*: n = 41 (1977), 28 (1981).

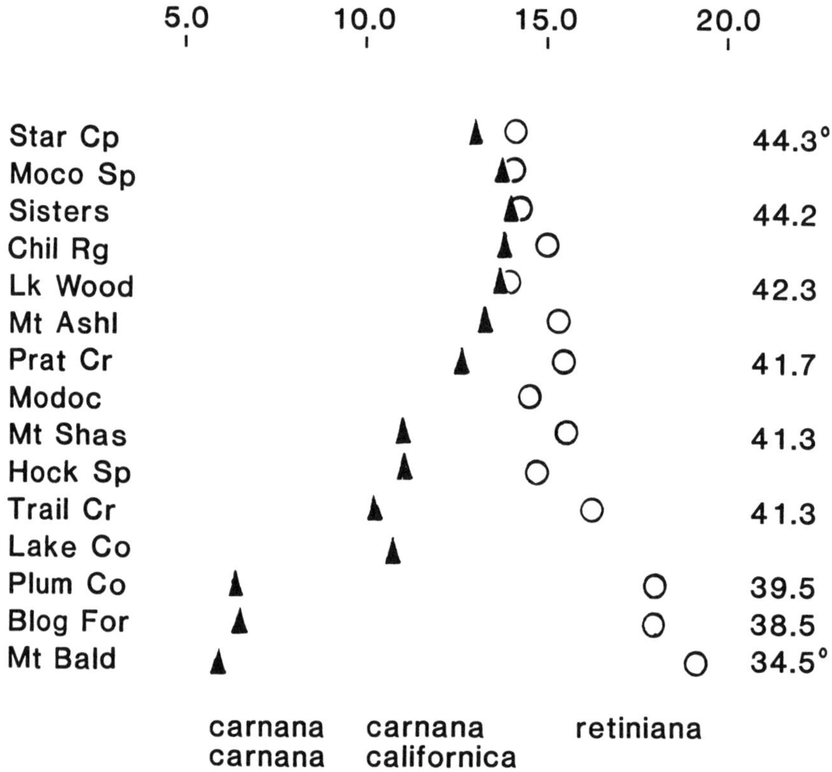

Fig. 2.6. Geographical variation in phenotype among populations of *C. carnana* (triangles) and *C. retiniana* (circles) from south to north (degree of latitude at right). Phenetic index (see text) on horizontal axis, from most *C. carnana*-like (5.0) to most *C. retiniana*-like (20.0). Symbols represent mean phenetic values for males attracted to *C. occidentalis* females or OXY synthetic lure (triangles), and for sympatric males attracted to *C. retiniana* females or RET-C lure (circles). The phenotypes converge to indistinguishable on an individual basis north of the Oregon border (42° N).

Appendix 2.1. Hostplant associations in natural populations of *Choristoneura* in California and Oregon, based on adults reared from field-collected larvae

Species	Lot no.	Locality	Hostplant
California:			
C. carnana carnana	80F38	6 km S Crystal Lake, Los Angeles Co.	*Pseudotsuga macrocarpa*
"	80F29	5 km NE Mt. Baldy, S. Bernardino Co.	"
"	81F106	"	"
"	82F18	Angelus Oaks, S. Berdo. Co.	"
C. carnana "Sierran population"	79F26, F33	Blodgett Forest, El Dorado Co.	*Pseudotsuga menziesii*
"	80F108	"	"
"	81F116	"	"
"	79F38	"	*Abies concolor*
C. carnana californica	79E42-43, E74	3 km NE Angwin, Napa Co.	*Pseudotsuga menziesii*
"	80E36	"	"
"	79E59, F17	1.5 km W Angwin	"
"	80E31	"	"
"	80E42	1.5 km SE Angwin	"
"	81E38	"	"
"	80F7	Mt. St. Helena, Napa Co.	"
"	79E80	1.5 km NE Cobb, Lake Co.	"
"	80E49, F12	"	"
"	80E55	2.5 km W Branscomb, Mendocino Co.	"
C. carnana "Cascades population"	74F60	Snowmans Summit, Siskiyou Co.	*Abies concolor*
"	80F43	Blue Mtn., Trinity Co.	*Pseudotsuga menziesii*
"	80F47	"	*Abies concolor*
"	80F49	Nelson Cr. Gap, Trinity Co.	*Pseudotsuga menziesii*
"	80F52	"	*Abies concolor*

C. carnana "Cascades population"	80F53	Scott Mtn., Trinity Co.	*Pseudotsuga menziesii*
"	80F58	Shadow Creek, Siskiyou Co.	"
"	80F70	6 km SW Etna, Siskiyou Co.	"
"	80F73	"	*Abies concolor*
"	82F76	Willow Cr., Siskiyou Co.	*Pseudotsuga menziesii*
C. r. retiniana	74F31	McBride Spr., Mt. Shasta, Siskiyou Co.	*Abies concolor*
"	74F56	So. Slope Mt. Shasta, 1,900m	"
"	74F58	Everitt Memorial, " 1,830m	"
"	74F44	9 mi N Bartle, Sisk. Co.	"
"	80F98	Rush Creek, Modoc Co.	"
"	80F105, G19	Cottonwood Spr., Modoc Co.	"
"	74G26, G44, G62	Manzanita Mtn., Modoc Co.	"
"	74G28, G46, G64	Snell Spr., Modoc Co.	"
"	70F83	2 mi NW Ft. Bidwell, Modoc Co.	"
C. r. retiniana	70F25, F68	Fandango Pass, Warner Mts., Modoc Co.	"
"	74G12, G30, G48	Fandango Peak, Warner Mts.	"
"	74F66	Mill Cr., Warner Mts.	"
"	74G16, G34, G52	Mill Cr., Warner Mts.	"
"	74F71	Shields Cr., Warner Mts.	"
"	74G20, G38, G56	Shields Cr., Warner Mts.	"
"	74F72, F73	Soup Cr., Warner Mts.	"
"	74G22, G40, G58	"	"
"	74G14, G32, G50	Benton Mdws., Warner Mts.	"

C. r. retiniana	74G18, G36, G54	So. Deep Cr., Warner Mts.	*Abies concolor*
"	74G24, G42, G60	Long Vy., Warner Mts., Lassen Co.	"
"	80F64	Hockaday Spr., Siskiyou Co.	"
"	80F93	Snowmans Summit, Siskiyou Co.	"
C. retiniana spaldingiana	70G26	Donner Lake, Nevada Co.	"
"	79G5	Ward Cr., 3 km SW Tahoe City, Placer Co.	*Abies magnifica*
"	79G10	6 km S Tahoe City, Placer Co.	"
"	80G17	"	"
"	79F30	Blodgett Forest, El Dorado Co.	*Abies concolor*
"	80F25	Tehachapi Mt., Kern Co.,	"
"	81F101, F104	"	"
"	82F15	"	"
"	81F113	Fir Ridge, Mt. Pinos, Ventura Co.	"
"	81F114	McGill Campgr., Mt. Pinos, Kern Co.	"
"	81F108	Blue Ridge Summit, Los Angeles Co.	"
Oregon:			
C. hybrid populations	70G13	25 air mi. E Medford, Jackson Co.	*Abies magnifica*
"	70G14	"	*Abies concolor/grandis*
	82H2-H7	Canyon Cr. Summit, Grant Co.	*Abies grandis*
"	82H8-H9	Delintment Lk., Harney Co.	"
"	82H10-H14	Moco Spr., Harney Co.	"
	California:		
"	80F83	Prather Cr., W of Macdoel, Siskiyou Co.	*Abies concolor/grandis*

C. lambertiana subretiniana	74G19, G55	So. Deep Cr., Warner Mts., Modoc Co.	*Abies concolor*
"	74G41	Soup Cr., Warner Mts.	"
"	70G22	Sagehen Cr., Nevada Co.,	"
	80G13	"	*Pinus contorta*
"	80G15	Donner Camp, Nevada Co.	"
"	79G1	Donner Summit, Nevada Co.	*Pinus albicaulis* (?)
"	80G18	6 km S Tahoe City, Placer Co.	*Abies magnifica*

3

Pheromone Chemistry and Response Behavior of Conifer-Feeding *Choristoneura* in the Western United States

Gary E. Daterman
Pacific Northwest Research Station
Forest Science Laboratory, Corvallis, Oregon

Laurence L. Sower
Pacific Northwest Research Station
Forest Science Laboratory, Corvallis, Oregon

Richard E. Stevens
Pacific Northwest Research Station
Forest Science Laboratory, Corvallis, Oregon

Douglas G. Fellin
Pacific Northwest Research Station
Forest Science Laboratory, Corvallis, Oregon

Abstract

Pheromone chemistry of *Choristoneura* in the western U. S. was studied to assist in clarifying differences among species and to produce an effective trapping method for studying geographic distributions and phenology. The taxa and source populations considered are: *C. retiniana* (Walsingham) (Bly, Oregon); *C. carnana californica* Powell (Angwin and Trinity Lake, California); *C. occidentalis* Freeman (Kamloops, B.C., Canada; McCall, Idaho; Missoula, Montana; Taos, New Mexico); *C. lambertiana subretiniana* Obraztsov (Cedarville, Calif.; Helena and Greenough, Mont.; Chemult, Ore.); *C. lambertiana ponderosana* Obraztsov (Lyons, Colorado).

Identifications of pheromone chemical components were determined by gas chromatography and mass spectrometer analysis, by collection of emissions from the surface of the pheromone glands or from a flow of air over calling females. Bioassays of candidate synthetic compounds were conducted at some of the same sites from which insect stocks had originated and at other localities believed to be inhabited by conspecific taxa; numbers of male moths captured were compared to those lured by conspecific virgin females. A previously developed synthetic lure for *C. occidentalis* (OXY: 92:8/E:Z-11-tdal) was effective from the onset of the study; synthetic blends for *C. retiniana* (RET: 94% 93:7/E:Z-11-tdac + 6% 92:8/E:Z-11-tdol or 90% 92:8/E:Z-11-tdac + 10% 90:10/E:Z-11-tdol) and for *C. lambertiana subretiniana* (SUBRET: 90% 60:40/E:Z-11-tdac + 10% 60:40/E:Z-11-tdol) were developed through a series of trials and reached effectiveness in 1981 and 1982. In each case, trapping in areas of California, Colorado, and Oregon from which female stocks originated lured conspecific males in numbers equivalent to or exceeding those of conspecific virgin females. Trapping results in Montana and in east central Oregon (Burns) were difficult to assess because large numbers of *C. occidentalis* males were attracted to RET and SUBRET lures, which raises questions about the degree of species integrity in these areas. Similar results were obtained with males attracted to RET and SUBRET baits and virgin females in south central Oregon (Klamath Lake area), where hybridization has been documented by previous studies. The recent recovery of a small but distinct aldehyde component in the pheromone emissions of an acetate-producing, pine-feeding *Choristoneura* from both Montana and Oregon provides strong evidence for natural hybridization of *C. occidentalis* and *C. lambertiana* in these areas.

INTRODUCTION

Because of the prevalence of polymorphism and other variation in western *Choristoneura* species, there is a clear need for more discerning characteristics as a basis for determining species and subspecies. Powell (1980) expressed the difficulties associated with classification of the western *Choristoneura.*

It is important to understand that names applied to both Abietoideae- and Pinidoideae-feeding *Choristoneura*, particularly in western North America, are little more than signs posted at irregular intervals on mosaics of differing populations. Any attempt to define species as reproductively isolated entities of comparable rank or genetic integrity on the basis of present knowledge will be artificial. Adults vary phenotypically between populations in average size, color, extent and kind of wing markings; larvae and pupae vary in intensity of integumental color; populations differ in whether or not genetic polymorphism is expressed and, if so, in the proportions of the polymorph phases.

Pheromone studies of conifer-feeding *Choristoneura* populations in the western United States were initiated as a means to assist in clarifying differences among species and to provide a sensitive trapping method for studying geographic distributions. The species and subspecies considered are *C. occidentalis* Freeman (the Western Spruce Budworm), *C. retiniana* (Walsingham) (the Modoc Budworm), *C. carnana carnana* (Barnes and Busck), *C. carnana californica* Powell, *C. lambertiana subretiniana* Obraztsov, and *C. lambertiana ponderosana* Obraztsov. Objectives were to develop synthetic pheromone lures that, by their selective attraction, would both discriminate among sympatric species at particular sites and provide a better means to determine the geographic distribution of the various species and subspecies.

STUDY PROCEDURES

Collections of *Choristoneura* populations were made from areas where particular species or subspecies were known to occur. For *C. occidentalis,* specimens were collected from: Taos, New Mexico; McCall, Idaho; Kamloops, British Columbia; Missoula, Montana; and Warm Springs, Oregon. In all cases, the collections were in forests of mixed conifers, but Douglas-fir, *Pseudotsuga menziesii,* was always abundant. Chemical analyses for *C. occidentalis* pheromones also were conducted on specimens from a non-diapause laboratory strain. *C. retiniana* for pheromone analysis were obtained near Bly,

Oregon, from a forest of predominantly White Fir, *Abies concolor*, intermixed with Lodgepole Pine, *Pinus contorta*, and Ponderosa Pine, *P. ponderosa*, but no Douglas-fir. There were no *C. carnana carnana* available for studies of pheromone chemistry, but the subspecies, *C. carnana californica*, was collected from Douglas-fir near Angwin, California, and the Trinity Lake area near Weaverville, California. The pine-feeder, *C. lambertiana ponderosana*, was obtained from the vicinity of Lyons, Colorado, in an area of abundant Ponderosa Pine; and *C. lambertiana subretiniana* was collected from a pure Lodgepole Pine stand in Oregon near Chemult, a mixed Lodgepole and White Fir stand in California near Cedarville, and areas of Ponderosa Pine, Douglas-fir, and Western Larch, *Larix occidentalis*, in Montana near Helena and Greenough.

Chemical identification of pheromone components was determined by gas chromatography and mass spectrometer analysis. The specifics of these analyses are described by Cory *et al.* (1982) and Silk *et al.* (1982) for *C. occidentalis* (OXY), and by Daterman *et al.* (1984) for *C. retiniana* (RET). Methods for pheromone identification of the western pine-feeding subspecies have followed a similar technique. In all cases, the preferred approach for analysis was to collect pheromone emissions from the surface of the insects' pheromone glands (hexane "wash" or "dips") or from a flow of air over the calling females ("effluvia").

Bioassays of the candidate pheromone compounds and blends of compounds were conducted at some of the same sites from which the insect stocks were originally collected. In other cases, the bioassays took place at locations where we were reasonably certain of the integrity of the species or subspecies inhabiting the site (Tables 3.1, 3.2). The bioassays consisted of trapping evaluations that compared numbers of male moths captured in traps baited with particular blends of synthetic chemicals, and with traps baited with unmated female moths of the particular species or subspecies being evaluated. The objective of these bioassays was to verify that the synthetic blend of compounds equaled or exceeded the attractiveness of the natural pheromone emitted by a live, unmated female moth.

At some locations, comparative trapping with live females of more than one species and synthetic lures known to be effective for two specific *Choristoneura* taxa was conducted to determine if any interspecific attraction occurred. The objective of this trapping was to evaluate the potential for hybridization that would be likely if cross-attraction of the different sympatric populations occurred. Comparative trapping for this purpose was conducted in areas harboring sympatric populations of *C. occidentalis* and *C. lambertiana subretiniana, C. occidentalis* and *C. retiniana,* and *C. retiniana* and *C. l. subretiniana.*

The pheromone chemistry of *C. occidentalis* was known at the outset of these studies, but the pheromones for *C. retiniana, C. l. subretiniana, C. l. ponderosana, C. carnana carnana,* and *C. c. californica* were either unknown or only partially identified. Consequently, in 1978, 1979, and 1980, the synthetic lures provided to Powell and De Benedictis (Part 2 this vol.) were inadequate for some of the study taxa. In 1981, the pheromone chemistry for the pine-feeders, *C. lambertiana subretiniana* and *C. l. ponderosana*, and the fir-feeding *C. retiniana*, had been identified (Daterman *et al.*

1984, and unpublished data). Consequently, beginning in the 1981 season, excellent synthetic lures were available for trapping all these *Choristoneura* taxa, and they were used for determining the extent of their geographic distributions (Powell and De Benedictis, Part 2 this vol.).

The synthetic lures provided prior to 1981 were partially effective in attracting some of the *Choristoneura* species and subspecies and other Lepidoptera. The content of the synthetic lures were as follows:

Choristoneura occidentalis (OXY):

1978-1981 = 92:8/E:Z-11- tetradecenal (tdal), (.1% pheromone lure concentration).

1982 = 92:8/E:Z-11-tdal, (.1 and .01% lure concentrations; .01% approximates release rate of an unmated female moth).

Choristoneura retiniana (RET):

1978, 1979 = E-11-tetradecenyl acetate (tdac), (.1% lure conc.) ("RET-A").

1980 = 95% E-11-tdac plus 5% E-11-tetradecenol (tdol), (.1% lure conc.) ("RET-B").

1981 = 94% 93:7/E:Z-11-tdac plus 6% 92:8/E:Z-11-tdol, (.1% lure conc.) ("RET-C").

1982 = 90% 92:8/E:Z-11-tdac plus 10% 90:10/E:Z-11-tdol, (.1% and .01% lure conc. with .01% lure approximating the release rate of an unmated female moth).

Choristoneura lambertiana subretiniana (SUBRET):

1978, 1979 = None.

1980 = 90% 78:22/Z:E-11-tdac plus 10% 78:22/Z:E-11-tdol, (.1% lure conc.) ("RET-A").

1981 = 90% 60:40/E:Z-11-tdac plus 10% 67:33/E:Z-11-tdol, (.1% lure conc.) ("RET-B").

1982 = 90% 60:40/E:Z-11-tdac plus 10% 60:40/E:Z-11-tdol, (.1% and .01% lure conc. with .01% approximating the release rate of an unmated female moth).

RESULTS AND DISCUSSION

Pheromone Chemistry

From the beginning of these studies, the OXY synthetic lures have been effective for trapping *Choristoneura occidentalis, C. c. carnana*, and *C. c. californica* males (Liebhold and Volney 1985, Powell and De Benedictis Part 2 this vol.). Subsequent chemical analyses of pheromone emissions from unmated *C. c. californica* females have confirmed that the sex pheromones of *C. occidentalis* and *C. c. californica* are very similar (Daterman *et al.* unpublished), as was indicated by field trapping with blends of synthetic pheromone compounds (Liebhold and Volney 1985, Powell and De Benedictis Part 2 this vol.). No *C. carnana carnana* females have been available for chemical analyses of the pheromone, but trapping data indicate that this subspecies shares a similar tdal pheromone with *C. occidentalis* and *C. carnana californica* (Powell and De Benedictis Part 2 this vol.).

In addition to the E/Z-11-tdals reported as the principal pheromone components produced by *C. occidentalis*, Silk *et al.* (1982) found E/Z-11-tdacs and E/Z-11-tdols in *C. occidentalis* female gland washes and effluvia. Alford and Silk (1984) reported that the addition of these "secondary" pheromone components enhanced laboratory bioassay responses of male *C. occidentalis*. Cory *et al.* (1982) also reported tdacs and tdols in female gland washes but determined that these components did not increase the attractiveness of trap lures in field conditions. Liebhold and Volney (1985) evaluated the effects of tdacs and tdols as additions to the basic 92:8/E:Z-11-tdal OXY lure and found no significant effects on trap captures of *C. occidentalis* males.

Nevertheless, significant questions have been raised concerning the possible roles of secondary pheromone compounds and related stimuli that might influence chemical communication within and among *Choristoneura* species (Silk *et al.* 1982, Alford and Silk 1984, Silk and Kuenen 1988, Grant 1987). For purposes of the studies reported in this volume, however, the synthetic tdal lure (OXY) was highly effective for capture of *C. occidentalis* and *C. carnana* males. For most study locations, the tdal lure was capable of selectively trapping these taxa where they occur in sympatric associations with other conifer-feeding *Choristoneura* species (Powell and De Benedictis Part 2 this vol.).

Choristoneura retiniana pheromone identification was sidetracked for a time because of misleading information from early field trapping studies with lures alleged to contain only E-11-tdac (Sanders *et al.* 1974). These early trials were successful in capturing large numbers of *C. retiniana* males, but later trapping efforts with new preparations of E-11-tdac were unsuccessful. Now that the *C. retiniana* pheromone is known (Daterman *et al.* 1984), logic suggests that the early successes were due to E-11-tdac lures that were contaminated with a small percentage of Z-11-tdac, which is an essential element for attraction.

Daterman *et al.* (1984) also demonstrated that additions of small quantities (10% of total active ingredients) of E/Z-11-tdols to synthetic lures significantly increased their attractiveness to *C. retiniana* males. For most locations, these lures have been effective for selective trapping of *C. retiniana* males, including in localities where the *C. retiniana* populations are sympatric with other species of *Choristoneura*. As reported elsewhere (Liebhold and Volney 1985, De Benedictis *et al.* Part 5 this vol.), there are locations where selective trapping is less effective for discriminatory capture of specific *Choristoneura* taxa. Furthermore, Sanders *et al.* (1977) reported that *C. retiniana* males, as well as other western *Choristoneura* species, responded with strong preference to their conspecific females, but they also demonstrated a slight degree of attraction to females of other *Choristoneura* species, in Oregon and northeastern California.

Choristoneura lambertiana subretiniana was the primary pine-feeding budworm studied for pheromone chemistry. As in the *C. retiniana* pheromone, the principal components of the *C. l. subretiniana* pheromone are the E/Z-11-tdacs, but the *C. l. subretiniana* female releases a larger proportion of the Z-isomer. The synthetic blend used for the studies reported in this volume (SUBRET) consisted of a 60:40/E:Z mixture of 11-tdacs with a

10% addition of the corresponding alcohols. For *C. l. ponderosana* (POND), a 70:30/E:Z mixture of tdacs was used, with a 15% addition of the corresponding alcohols. These blends were bioassayed as lures in adhesive traps in Montana, Oregon, and Colorado. Where available, the synthetic lures were compared to the attraction by *C. l. ponderosana* and *C. l. subretiniana* females placed in screen cages inside the adhesive traps.

Field Bioassays

Field comparisons of live female attraction versus synthetic lures indicated the synthetics were at least as effective in attracting males as were the conspecific females. For both *Choristoneura occidentalis* and *C. retiniana*, the synthetic lures captured approximately twice as many males as the traps baited with females (Cory *et al.* 1982, Daterman *et al.* 1984), when the synthetic lures contained 0.01% pheromone, a concentration approximating the natural female release rate.

For *C. lambertiana subretiniana* and *C. l. ponderosana*, similar responses occurred at Oregon and Colorado trapping sites, with the insects' respective synthetic lures capturing at least twice as many male moths as the traps baited with conspecific females (Table 3.3). At the Chemult, Oregon, location the *C. l. "subretiniana"* females from Montana used as baits and the *C. l. ponderosana* females from Colorado proved far less attractive to *C. l. subretiniana* males than the *C. l. subretiniana* females from the Chemult location (Table 3.3). In contrast, a subsequent field bioassay in the vicinity of Cedarville, California, showed the Montana *C. l. "subretiniana"* females to be equally attractive as local *C. l. subretiniana* females (Daterman, unpubl. data). Possibly the Montana *C. l. subretiniana* and Colorado *C. l. ponderosana* females used at the Chemult location were in poor condition, or their diel pheromone "calling" period may be asynchronous with the time of *C. l. subretiniana* male flight in Oregon.

The results of field trapping bioassays in Montana were not as clear because *Choristoneura occidentalis* males were captured in the traps baited with RET and SUBRET synthetic lures. Large numbers of captured specimens in the traps made determinations of *C. occidentalis* and *C. l. "subretiniana"* males difficult because many of the specimens were damaged. Relatively dense local populations of *C. occidentalis* were present in the area of sympatric populations, and the large numbers of moths probably contributed to the problem of cross-attraction to the SUBRET lures. These results, however, raised a question of the degree of species integrity in this area. Clearly, the *C. occidentalis* males were not restrained from entering traps from which synthetic pheromone of another species was being released.

Interspecific Attraction

In areas of sympatric *Choristoneura* populations, the respective species usually maintain their integrity in response to their particular pheromone blends, whether it be represented by synthetic or natural lures. In some areas, however, the selection by males of the appropriate lure breaks down, and interspecific response occurs. Because such behavior could lead directly to interspecific mating and hybrids of *Choristoneura*, we record some particular occurrences of interspecific pheromone responses that may provide further insight concerning the relative proclivities of these taxa to hybridize under natural conditions. Sanders *et al.* (1977) reported the relative ease of obtaining interspecific matings and hybrid progeny by caging adults of different species together in the laboratory.

Choristoneura occidentalis and *C. l. lambertiana subretiniana* populations occur sympatrically over wide areas of western forests, particularly in Oregon, Washington, Montana, and Idaho. Recent chemical analyses of *C. l. subretiniana* female pheromone washes indicated a small but distinct complement of aldehyde present, along with the expected acetate and alcohol components. These aldehyde-containing *C. l. subretiniana* specimens were collected from mixed Douglas-fir/ Ponderosa Pine forests near Burns, Oregon, and a similar mixed-species forest near Greenough, Montana, both areas having substantial populations of *C. occidentalis*. The quantities of aldehyde present in gland washes were always 10% or less of the total pheromone components (Daterman *et al.* unpubl. data).

A small complement of aldehyde was added to the standard synthetic SUBRET lure to determine if responses from *C. occidentalis* or *C. l. subretiniana* males might be indicative of how interspecific attraction and hybridization could occur in natural conditions. Two SUBRET lures were prepared, one containing a 5% proportion of aldehyde, and the other consisting of the standard SUBRET lure. Both contained a total of 0.01% total synthetic pheromone to approximate the release rate of *Choristoneura* female. Ten traps of each were exposed for four nights in the Burns, Oregon, area. The traps containing the aldehyde component captured 42 male *C. occidentalis*, and the standard SUBRET synthetic lure captured 5 *C. occidentalis* males. Clearly, *C. occidentalis* males were attracted to the SUBRET lure with the added complement of tdal, even though 95% of the lure components consisted of tdacs and tdols; the presence of a much greater quantity of the SUBRET pheromone components did not deter the *C. occidentalis* males. This is consistent with the findings of Sanders et al. (1977) showing that *C. occidentalis* males readily responded to traps baited with a *C. occidentalis* female plus a female of another *Choristoneura* species. Significantly, the lure containing no aldehyde also captured a few *C. occidentalis* males, indicating that populations that have been assigned to this species in central Oregon represent a hybrid zone in which males are not discriminating when seeking potential mates. These findings suggest that a percentage of *C. occidentalis* versus *C. l. subretiniana* hybridization could occur from *C. occidentalis* males responding to *C. l. subretiniana* females, and that if a hybrid female capable of producing some tdal in her pheromone emissions should occur, that *C. occidentalis* male response would be even stronger.

Choristoneura retiniana and *C. lambertiana subretiniana* populations occur sympatrically in the Warner Mountains of northeastern California and south-central Oregon. The predominant forest composition in these areas is White Fir, Lodgepole Pine, and Ponderosa Pine. In recent tests to evaluate different blends of synthetic chemicals as lures for *C. l. subretiniana* males, substantial numbers of *C. retiniana* males were also captured (Daterman unpubl. data). Further, traps baited with live female *C. l. subretiniana* also captured *C. retiniana* males. This result occurred in an area of 90% or more Lodgepole Pine, with only occasional White Fir. Presumably there was not a local abundance of *C. retiniana* females available to *C. retiniana* males flying in the area. This suggests a general tendency for *Choristoneura* males in central and southern Oregon to respond to females of other *Choristoneura* species when a conspecific pheromone source is scarce or nonexistent in the local area. Such circumstances might occur at the end of a flight season when most females of a species have completed mating and are no longer releasing pheromone, and particularly where one species of a sympatric association is in a high density, which serves to flood an area with males.

Furthermore, the numbers of "interspecific" captures of *C. retiniana* and *C. l. subretiniana* increased as the synthetic blends of E/Z tdac isomers changed in the direction of the other species' authentic blend (Daterman unpubl. data). If some variation in natural blends occurs among individual females, this would contribute to the occurrence of natural hybridization.

The occurrence and potential for *C. retiniana* versus *C. occidentalis* interspecific attraction and hybridization in Oregon has been discussed in prior papers (Sanders *et al.* 1977, Liebhold and Volney 1984, 1985, Liebhold *et al.* 1984). In our studies, captures in traps baited with live females were generally specific for the conspecific males, but in areas of hybridization of sympatric association of *C. retiniana* and *C. occidentalis* populations, such as Chiloquin, Oregon, there is a small percentage of cross-attraction, particularly by *C. occidentalis* males to *C. retiniana* females. This is consistent with the observation by Sanders *et al.* (1977) that *C. occidentalis* males are not deterred by the presence of a different species of female in the same trap as the *C. occidentalis* female. Sanders *et al.* also reported the same response by *C. retiniana* males when *C. occidentalis* and *C. retiniana* females were jointly used as trap lures. Liebhold and Volney (1985), however, found that the addition of tdals to a standard synthetic *C. retiniana* lure significantly decreased attraction of *C. retiniana* males.

CONCLUSIONS

The results of these studies indicate that synthetic blends of the appropriate pheromone chemicals can be used to selectively capture *Choristoneura* males in areas of sympatric distribution with related species. In some cases, such as in the hybridization zone of central and southern Oregon, there is a degree of interspecific attraction among the conifer-feeding *Choristoneura*, even though it is relatively low in comparison to the preferred conspecific responses of the respective species studied in most regions. The interspecific attraction, however, almost certainly results in some instances of natural hybridization in the field. The recent recovery of a small but distinct aldehyde component in the pheromone emissions of an acetate-producing, pine-feeding *Choristoneura* from both Oregon and Montana collections provides strong evidence for natural hybridization between the Western Spruce Budworm, *C. occidentalis*, and the pine-feeding *C. lambertiana [C. l]. subretiniana* .

Because males of *C. occidentalis* seem to be especially aggressive in entering traps containing live females of other species as lures, particularly when their conspecific pheromone is present, we propose that they are more likely to initiate natural hybridization among the western species. A scenario that especially lends itself to the occurrence of hybridization is a sympatric association of *Choristoneura* species in which unmated females of one of the species are severely limited for temporal or other reasons. In such cases, we suggest that interspecific attraction and propensity for hybridization is likely.

LITERATURE CITED

Alford, A.R., and P.J. Silk. 1984. Behavioral effects of secondary components of sex pheromone of western spruce budworm. J. Chem. Ecol. 10:265-270.

Cory, H.T., G.E. Daterman, G.D. Daves, Jr., L.L. Sower, R.F. Shepherd, and C.J. Sanders. 1982. Chemistry and field evaluation of the sex pheromone of western spruce budworm, *Choristoneura occidentalis* Freeman. J. Chem. Ecol. 8:339-350.

Daterman, G.E., H.T. Cory, L.L. Sower, and G.D. Daves, Jr. 1984. Sex pheromone of a conifer-feeding budworm, *Choristoneura retiniana* Walsingham. J. Chem. Ecol. 10:153-160.

De Benedictis, J.A., A.M. Liebhold and J.A. Powell. 1995. Studies of pheromone attraction and inheritance, using virgin females of *Choristoneura* species, in the central Sierra Nevada, California. Univ. Calif. Publ. Entomol. (Part 5 this vol.).

Grant, G.G. 1987. Copulatory behavior of spruce budworm, *Choristoneura fumiferana* (Lepidoptera: Tortricidae): Experimental analysis of the role of the sex pheromone and associated stimuli. Ann. Entomol. Soc. Amer. 80:78-88.

Liebhold, A.M., and W.J.A. Volney. 1984. Effect of temporal factors on reproductive isolation between *Choristoneura occidentalis* and *C. retiniana* (Lepidoptera: Tortricidae). Canad. Entomol. 116:991-1005.

Liebhold, A.M., and W.J.A. Volney. 1985. Effects of attractant composition and release rate on attraction of male *Choristoneura retiniana, C. occidentalis* , and *C. carnana* (Lepidoptera: Tortricidae). Canad. Entomol. 117:447-457.

Liebhold, A.M., W.J.A. Volney, and W.E. Waters. 1984. Evaluation of cross-attraction between sympatric *Choristoneura occidentalis* and *C. retiniana* (Lepidoptera: Tortricidae) populations in south-central Oregon. Canad. Entomol. 116:827-840.

Powell, J.A. 1980. Nomenclature of nearctic conifer-feeding *Choristoneura* (Lepidoptera: Tortricidae): Historical review and present status. USDA-For. Serv., Pacific NW For. and Range Expt. Sta., Gen. Tech. Report PNW-100; 18 pp.

Powell, J.A., and J.A. De Benedictis. 1995. Biological relationships: Host tree preferences and isolation by pheromones among allopatric and sympatric populations of western *Choristoneura* . Univ. Calif. Publ. Entomol. (Part 2 this vol.).

Sanders, C.J., G.E. Daterman, and T.J. Ennis. 1977. Sex pheromone responses of *Choristoneura* spp. and their hybrids (Lepidoptera: Tortricidae). Canad. Entomol. 109:1203-1220.

Sanders, C.J., G.E. Daterman, R.F. Shepherd, and H. Cereske. 1974. Sex attractants for two species of western spruce budworm, *Choristoneura biennis* and *C. viridis* (Lepidoptera: Tortricidae). Canad. Entomol. 106:157-159.

Silk, P.J., and L.P.S. Kuenen. 1988. Sex pheromones and behavioral biology of the coniferophagous *Choristoneura*. Ann. Rev. Entomol. 33:83-101.

Silk, P.J., C.J. Wiesner, S.H. Tan, R.J. Ross, and G.G. Grant. 1982. Sex pheromone chemistry of the western spruce budworm *Choristoneura occidentalis* Freeman. J. Chem. Ecol. 8:351-362.

Snedecor, G.W. 1956. Statistical methods applied to experiments in agriculture and biology. Iowa St. Coll. Press, Ames; 534 pp.

Table 3.1. Collections of *Choristoneura* from various trapping locations, using synthetic RET lures and/or live female *C. retiniana*

Location	Synthetic Lure[1]	Live Female Lure
Chiloquin, Ore.	+	+
Lakeview, Ore.	+	+
Bly, Ore.	+	+
Fort Klamath, Ore.	+	+
Cedarville, Calif.	+	+
Burns, Ore.	+	NT
Enterprise, Ore.	-	-
Ellensberg, Wash.	NT	+ (?)
Manitou, Colo.	+ (?)	NT

1. + = *Choristoneura* males captured; - = no male captures; NT = no test for the lure at a particular location; (?) = identities of captured males uncertain, probably *C. occidentalis*.

Table 3.2. Collections of *Choristoneura lambertiana* (*s. lat.*) from various trapping locations using synthetic SUBRET and/or live females as lures

Location	Synthetic Lure[1]	Live Female Lure
Chemult, Ore.	+	+
Chiloquin, Ore.	+	+
Lakeview, Ore.	NT	+
LaGrande, Ore.	+	NT
Burns, Ore.	+	+
Ellensburg, Wash.	NT	+
Goldendale, Wash.	+	NT
Helena, Mont.	+	+
Greenough, Mont.	+	+
Cedarville, Calif.	+	+
Manitou, Colo.	+ (?)	NT
Poudre, Colo.	+ (?)	NT
St. Vrain, Colo.	+ (?)	NT

1. + = *C. l. subretiniana* males; NT = no test; (?) = "*subretiniana*" present in *ponderosana* population, represented by males that were more mottled than typical *ponderosana* phenotype, as judged by Stevens.

Table 3.3. Average numbers of male *Choristoneura lambertiana subretiniana* (at Chemult, Oregon) and *C. l. ponderosana* (at Lyons, Colorado) captured in traps baited with synthetic pheromone blends or unmated females.

Trap Lure	Oregon[1]	Colorado
SUBRET synthetic	70[a]	25[a]
POND synthetic	64[a]	26[a]
Oregon female *subretiniana*	33[b]	No test
Colorado female *ponderosana*	2[c]	5[b]
Montana female "*subretiniana*"	2[c]	No test
Trap replications	15	10

1. Within each column, means followed by same letter do not differ (P=.05), according to Tukey's Test (Snedecor 1956).

4

Phenetic Studies of Spruce Budworm Populations (*Choristoneura* Species)

John A. De Benedictis
Entomology Department
University of California, Davis, California

Abstract

Eight morphological characters were used in phenetic studies of 384 males from 40 western U.S. populations of conifer-feeding *Choristoneura* species. Within each of three species-groups based on pheromone attraction, larval host, or resemblance to a distinct population, characters were generally clinally distributed among coastal populations and randomly distributed among interior populations. Using Euclidean distance between principal components, a nearest neighbor of every population was assigned *a priori* to the same subspecies except for a population in the "retiniana" species-group from a putative hybrid zone which linked with *C. occidentalis* populations. Most subspecies linked with another of the same species or species-group, except for pine-feeding populations from coastal Oregon and Washington. Discriminant function analysis assigned only 38.3% of the individuals each to its own population, but 73.0% and 75.2% to the same *a priori* subspecific and specific classification of its population, respectively. The analyses of the characters considered here indicate that *C. spaldingiana* should be considered a subspecies of *C. retiniana*, that heavily marked pine-feeding populations from the western Rockies can be included in *C. lambertiana subretinana*, and that Sierra Nevada populations of *C. carnana* are more similar to *californica* than the nominate subspecies. The relationships of the coastal Oregon and Washington pine-feeding populations and of *carnana*-like specimens from Arizona to other *Choristoneura* need clarification.

INTRODUCTION

In his review of the spruce budworms, the conifer-feeding members of the genus *Choristoneura* Lederer, Powell (1980) summarized the taxonomic history of the group and presented a revised treatment which otherwise would have appeared in its revised status in the checklist of North American Lepidoptera (Hodges *et al.* 1983) without prior explanation. In these two works, 9 of the 16 species-group taxa from North America were recognized as species, and 4 were recognized as subspecies. The other names were placed in synonymy.

Powell (1980) noted that nomenclatural constraints require that taxa be classified at various taxonomic levels which do not reflect the biological complexity of the group. Species of spruce budworms exhibit various degrees of isolation and intermediacy which vary geographically and, perhaps, temporally. Hence, the application of the available names depends upon one's interpretation of the biological relationships and one's definition of a species.

Faced with such taxonomic uncertainty, I undertook a study of 40 *Choristoneura* populations from 27 localities in the United States west of the continental divide to identify morphological characters which might be useful in distinguishing among species and subspecies, to assess the distinctiveness and variability of the named forms, and to examine phenetic relationships among them.

Most of the named taxa in this group were described on the basis of wing pattern and coloration. Some have distinctive phenotypes; others do not. Later descriptions also included perceived differences in genitalia and in larval coloration and hosts. However, adults of some populations are extremely variable, and larval hosts and coloration are often unknown, so identification can be exceptionally difficult. Consequently, I sought some other criterion for *a priori* classification of budworm populations.

A solution to this problem arose with the development of synthetic chemical attractants based on the females' sex pheromones. In field tests from 1978 to 1982, Powell and I attracted male spruce budworms to one of three chemical compounds formulated by G. E. Daterman (Daterman *et al.* Part 3 this vol.). The compounds have not been tested on some important populations, but in most regions they can be used to separate spruce budworms into three groups. I used this differential chemical attraction to define three major groups of this study which I refer to as "species-groups" to avoid confusing them with the species complexes of Powell (1980). Each of these species-groups may be composed of single species or groups of related species having identical or very similar sex pheromones.

Most of the 40 study populations were assigned to species-groups on the basis of chemical attraction. Populations whose chemical attraction was not tested were excluded from this study, unless they closely resembled a distinctive population whose attraction was known. Such populations were assigned to the species-group of the similar population.

By selecting sites for testing synthetic pheromone attraction where taxa were described or where populations resembling distinctive named taxa occur, the western subspecies recognized by Powell (1980) could be assigned to species-groups. Other populations whose identities were uncertain still could be placed into one of the groups once chemical attraction was determined.

The "Carnana" Group

Males of the "carnana" group are drawn to the "occidentalis" (OXY) synthetic pheromone which initially was used to attract *Choristoneura fumiferana* (Clemens), the Spruce Budworm of eastern North America, and subsequently was found to attract *C. occidentalis* Freeman (Sanders *et al.* 1974). The group is named after the oldest valid name of western moths attracted to that compound and is composed of the "carnana" and "occidentalis" species complexes of Powell (1980). It includes populations whose preferred, if not sole larval host is in the genus *Pseudotsuga* Carrière. Included are *C. carnana carnana* (Barnes and Busck), *C. carnana californica* Powell, and *C. occidentalis*.

The name *C. occidentalis* was given to a population of moths from eastern Washington (Freeman 1967), but has been applied to many western budworm populations associated with the preferred host, Douglas-fir, *Pseudotsuga menziesii* (Mirbel) Franco, including some hybrid populations. It is not known whether moths at the type locality are drawn only to the OXY attractant or to others.

In the absence of such information, I have applied the name *C. occidentalis* to all populations drawn to the OXY attractant which are not *C. carnana* (*sens. lat.*). Included among populations of *C. occidentalis* are moths from the vicinity of Flagstaff, Arizona, many of which more closely resemble *C. carnana* than *C. occidentalis*. I have excluded from this study several populations of putative *C. occidentalis* from the Pacific Northwest because of uncertainty about their chemical attraction.

Other species drawn to the OXY attractant, but not examined because they occur outside of the western United States, are *C. biennis* Freeman and *C. fumiferana* (Sanders *et al.* 1974). Sixteen of the 40 study populations are in the "carnana" group.

The "Lambertiana" Group

Moths attracted to the "subretiniana" (SUBRET) synthetic pheromone, which was based on the female sex pheromone of a laboratory colony of *C. lambertiana subretiniana* Obraztsov (G. E. Daterman, pers. comm.), are in the "lambertiana" group. This group includes *C. lambertiana* (Busck) and populations that are considered to comprise its subspecies, *C. l. subretiniana* and *C. l. ponderosana* Obraztsov, all of which prefer species of *Pinus* Linnaeus as the larval host.

Although their chemical attraction has not been tested, I also have included two populations from coastal Oregon and Washington in this species-group because of their association with pine.

The SUBRET pheromone also has not been tested on *C. l. lambertiana*, but this taxon is more similar phenotypically to *C. l. subretiniana* than are the coastal pine-feeding populations, so its assignment to this group is less speculative. I examined 12 populations in the "lambertiana" group.

The "Retiniana" Group

The "retiniana" group includes populations drawn to the "retiniana" (RET) attractant. The compound is based on the female sex pheromone of a laboratory colony of *Abies*-feeding *Choristoneura* originally collected near Bly, Oregon (Sanders *et al.* 1974, G. E. Daterman, pers. comm.), that Powell (1980) assigned to *C. retiniana* (Walsingham). I have restricted the name *C. retiniana* to populations from Oregon and northern California, at least some of which hybridize with moths from the other species-groups, especially when population densities are high.

In the Great Basin and in California south of the Cascades, moths in the "retiniana" group apparently never outbreed and, consequently, are less variable. These less variable populations are referred to as *C. spaldingiana* Obraztsov, although subspecific application of this name may be more appropriate. Of uncertain status when Powell (1980) reviewed spruce budworm taxonomy, moths of the distinctive *spaldingiana* phenotype were drawn to RET attractant near the type locality, Provo, Utah. Both species (or subspecies) are associated with White Fir, *Abies concolor* (Gordon and Glendinning) Lindley, and related firs. Their larvae appear green rather than brown with white pinacula as is characteristic of the larvae of the other species-groups. Twelve populations were examined in the "retiniana" group.

Except where it is an obvious consideration, I have ignored temporal variation in these studies. However, phenotypes and biological traits, such as population densities and hybridization, seem to vary from time to time as well as from place to place.

I also have ignored a possible nomenclatural problem concerning *C. occidentalis*. Volney *et al.* (1984) determined that in south-central Oregon, populations from two of the species-groups hybridize, unlike their behavior in many parts of the west. Using chemical attractants and by collecting larvae and pupae, Powell and I determined that the region of hybridization extends at least to northern Oregon east of the Cascade Range (Powell and De Benedictis Part 2 this vol.). It is possible that the hybrid zone includes the type locality of *C. occidentalis*, where chemical attraction has not been tested. A bioassay of the pheromone produced by females from there was inconclusive (De Benedictis *et al.* Part 5 this vol.). Hybrids cannot be differentiated reliably from *C. occidentalis*, so it is possible that the name was given to a hybrid population. If so, the name would have no taxonomic standing according to Article 23h of the International Code on Zoological Nomenclature.

MATERIALS AND METHODS

Study Populations

Powell and I collected all or some of the moths in 20 of the 40 populations examined in this study during field studies from 1978 to 1982. The two populations (C1 and R1) from the hybrid zone were collected using funnel traps baited with synthetic pheromones. At most other collecting sites, we collected moths in sticky traps baited with chemicals or caged virgin females. These moths were usually unsuitable for this study, but they enabled us to determine the range of phenotypes attracted to each pheromone. Except in the hybrid zone, the moths attracted to each pheromone were distinguishable from those attracted to the other pheromones so that undamaged specimens from most collecting sites could be assigned to one of the three species-groups. Other individuals and populations, samples of which were in the Essig Museum of Entomology, University of California, Berkeley, or borrowed from other institutions, were added to the study set if chemical attraction at the collection site was known or if they resembled a distinctive population whose attractant was known. Populations whose species-group assignment was uncertain were not included in this study.

Usually ten males were selected at random from a larger series to represent the population, with preference given to males with intact abdomens, legs, and wings. In some cases, fewer than ten males were available, but the population was included because of its locality or its distinctive phenotype. A total of 384 males represented the 40 populations (Appendix 4.1). Only males were used to avoid differences due to primary and secondary sexual characteristics, and because only males are drawn to the attractants.

The populations and their collection sites are shown in Figure 4.1. All populations of the "carnana" group from outside of California are referred to as *Choristoneura occidentalis*. Most other populations were assigned one of the names recognized by Powell (1980) as its *a priori* classification if they resembled a named species or subspecies in the species-group of the population. The remaining populations are referred to by some other designation as discussed in the descriptions of the species-groups which follow.

In the "carnana" group, *C. carnana carnana* is represented by population C6 from Mt. San Antonio (formerly Mt. Baldy), San Bernardino Co., California, the probable type locality. The symbol of this subspecies is a closed (filled) circle in the figures. Half-filled circles indicate other populations of *C. carnana* (C2-C5). I have referred to all of them *a priori* as *C. carnana californica*, even though Sierran populations more closely resemble the nominate subspecies than *californica* by facies. This arbitrary decision can be supported on the basis of preferred larval host. In the "carnana" group, only *C. c. carnana* (*sens. str.*) feeds upon *Pseudotsuga macrocarpa* (Vasey) Mayr, the only available species of *Pseudotsuga* where *carnana* (*sens. str.*) is found. *Pseudotsuga macrocarpa* is replaced by *P. menziesii* where the other populations in the "carnana" group occur.

The remainder of the "carnana" group (C1 and C7-C16), symbolized by open circles, are treated as *Choristoneura occidentalis*. I have applied this name to several atypical populations drawn to the "occidentalis" attractant. One such population is C11, collected in

1980 in the Schell Creek Range of White Pine Co., Nevada. Individuals are smaller than those of most other *C. occidentalis* populations; males are brown and females are rust, with little color variation; and every moth is finely mottled, often over the entire forewing. In 1980 we collected similar moths at 2,600 m in the nearby Snake Range, and in 1978 and 1981 Powell and I collected moths (C12) at 2,300 m in the Snake Range that resemble the more typical, larger and less mottled *C. occidentalis* phenotype.

Population C16 from 25 km southwest of Flagstaff, Arizona, is composed of small moths, many of which closely resemble *C. c. carnana* with slight differences in coloration and wing banding. Other moths of this population have a browner, somewhat mottled ground color that reaches the distal portion of the wing, unlike those that resemble *C. carnana*. A lesser proportion of the moths of nearby population C15 from north of Flagstaff resemble *C. c. carnana*. Many others are similar to the mottled moths of population C16, but slightly larger and darker. Size is bimodal in the available sample, with some moths still larger and darker like those of most other populations of *C. occidentalis*.

Individuals of population C1 from Grant County, Oregon, display sizes, markings, and coloration similar to typical *C. occidentalis*, but concurrent collections of immatures and chemical attraction tests indicate that some moths of the "carnana" group hybridized with those of the "retiniana" group there. C1 and the sympatric *C. retiniana* population (R1) are represented by moths drawn to the chemical attractants (C1 to the OXY bait and R1 to the RET bait). I refer to C1 and R1 as hybridizing populations and have placed a star in their symbols in the figures to differentiate them.

Of the 12 populations in the "lambertiana" group, *Choristoneura lambertiana lambertiana* is represented only by paratypes from Ashland, Oregon (L3) and is symbolized by a small closed circle within a triangle. The type series, although reared, is generally in poor condition. All but one male in the available sample lacked its abdomen or many of the legs.

Half-filled triangles symbolize *C. l. subretiniana* (L4-L6) and a similar, less mottled, unnamed group of populations from the northern Rocky Mountains that I call *C. l.* near *subretiniana* (L7 and L8). The left half of the triangle is filled in the symbols for the latter populations, while the right half is filled in symbols for *C. l. subretiniana*. *Choristoneura l. ponderosana* (L9-L12) is symbolized by filled triangles. I include populations from the Great Basin, many specimens of which differ from typical *C. l. ponderosana* by having rust-red blotches on their forewings.

The open triangles represent coastal pine-feeding populations (L1 and L2) whose tentative assignment to this species-group is based upon larval hostplant. L2 was not reared, but Lodgepole Pine is the dominant conifer at the collection site, and the moths are phenotypically very similar to L1, which was reared from Lodgepole Pine.

Open squares designate *Choristoneura retiniana* (R1-R3), restricted here to populations drawn to the RET attractant in Oregon and California from the southern end of the Cascade Range northward. The star in the symbol for population R1 indicates that it is hybridizing with the sympatric population of the "carnana" group. The remaining nine populations in the "retiniana" group are *C. spaldingiana* (R4-R12), symbolized by filled squares.

Except in the hybridizing population R1, populations of *C. retiniana* and *C. spaldingiana* include a form that Obraztsov (1962) described as *C. lindseyana*. This distinctive form has tan forewings, often with faint, brown longitudinal streaks. Proportions of this form among populations of *C. retiniana* and *C. spaldingiana* are listed in Appendix 4.2, and collection data for all study populations are listed in Appendix 4.1. Powell and De Benedictis (Part 7 this vol.) give diagnoses of the subspecies and discuss variant populations.

Morphological Characters

Eight morphological characters were examined (Appendix 4.2). A few others were considered but rejected because of high correlation with wing length, an indicator of size (Miller 1977). Forewing lengths and widths were measured with an ocular micrometer in a Bausch and Lomb binocular dissecting scope. The length was measured parallel to the costa from base to most extreme wing margin and excludes the fringe. The width was measured perpendicular to the dorsal margin at the apex of the discal cell.

Genitalic preparations were mounted on slides with the valvae spread and the uncus flattened. After drying, the slide was placed in a Bausch and Lomb microprojector, and the image of the genitalia was projected onto white paper at a fixed distance so that the focused image was enlarged about 22X. The minimum width of the uncus constriction and the maximum widths of the basal and apical portions of the projected image of the uncus (Fig. 4.2) were measured to 0.1 mm with vernier calipers.

A midleg and hindleg (the right, if present) were removed, soaked in a cold solution of 20% potassium hydroxide for about 24 hours then descaled in water by gentle abrasion. The legs were passed through dehydrating solutions, clearing agents, stain, and fixative with the genitalic preparation with which they were mounted. Slide numbers later were used as the individual specimen numbers in the statistical analyses. Numbers of basitarsal spines on each leg (Fig. 4.3) and cornuti in the aedeagus were counted under magnification of an Olympus phase contrast microscope.

Statistical Analyses

Descriptive statistics of the characters are listed in Appendix 4.2. Correlations and covariances among the characters are given in Table 4.1. Each character was analyzed for equality of population means using one-way analysis of variance (ANOVA). Population L3 is omitted, because there was only one genitalic preparation and two measurements of the other characters.

The simultaneous testing procedure (STP) of Gabriel (1964), as employed by Power (1970), determined which groups of populations have 95% probabilities of equal means for the eight characters. Inasmuch as STP relies upon the estimated overall variance (the within mean square of ANOVA), and ANOVA assumes that group (population) variances are equal, the equality of the population variances was tested using the Lev1:Med test (Conover *et al.* 1981).

In Figures 4.4-4.11, target symbols as utilized by Power (1970) replace the population symbols to illustrate five size categories for each of the eight characters based upon means. Empty symbols depict populations whose means are in the lowest fifth of the range of means, one filled quadrant denotes populations in the next lowest fifth, etc. Again, population L3 is omitted.

Using seven of the characters as variables, canonical discriminant function analysis (d.f.) and principal component analysis (p.c.) were performed. To make more individuals available for these analyses by eliminating a source of missing data, the cornuti counts were not used, because only the extreme populations were significantly different from one another. Cornuti counts were not possible when the aedeagus was lost or damaged, or if the cornuti had detached and their points of attachment were not clearly visible.

Discriminant function often is used to classify unknown individuals by comparing them to individuals of known classification. In this study, however, d.f. was used only to determine whether and how well the seven characters classified the individuals of the 40 study populations. Using d.f. to discriminate among the populations, a series of canonical variables that maximize the distances between the centroids of the study populations was calculated, a centroid being the average value of the discriminant scores of the individuals comprising each population. The Euclidean distances of the individuals from the centroids were calculated, and each was assigned *a posteriori* to the closest population with a probability related to its distance from the centroid of that population.

I used d.f. to see how many individuals in the study sample were assigned, each to its own population, and to the same *a priori* subspecific, specific, and species-group classification as its own population, and to analyze incorrect classifications. Also, I examined the distances between population centroids. Presumably, closely related populations will be closer to one another than to other populations.

Principal component analysis calculates a series of independent canonical variables, principal components, of which the first accounts for the greatest proportion of the total variation in the individuals, the second accounts for the greatest proportion of the remaining variation, etc. Unlike d.f., the *a priori* classification is not considered in calculating the principal components, so scattergrams of the principal component scores are useful in determining "natural" clusters of individuals or populations. Each principal component is a linear equation, and the coefficient of each character in the equation indicates the relative contribution of the character to the principal component. If one or more principal components successfully segregate the populations, the coefficients of those principal components reveal the important diagnostic characters.

RESULTS

All eight characters are positively correlated with one another (Table 4.1). Except where noted, the correlations are all significantly greater than zero ($r>0$).

Based upon analyses of variance (ANOVA), the null hypotheses, that population means of the morphological characters are equal, are rejected in all instances (Table 4.2).

Similarly, the null hypotheses that population variances are equal also are rejected using Lev1:Med tests (Table 4.2). Because there were only one or two data points, population L3 was omitted from these tests.

Figures 4.4-4.11 list populations and their means arranged from largest to smallest adjacent to the map. The vertical bars to the right of the population codes indicate sets of populations whose means are not significantly different from one another by the simultaneous testing procedure (STP) at α=0.05. However, because the population variances are not equal, the estimated overall variances (the within mean squares from ANOVA) may be somewhat low. If so, the 95% confidence intervals should be extended slightly, possibly to include other populations in some cases.

For all characters, no sets of populations completely segregate into discrete, non-overlapping, non-significant groups. However, for most characters, the non-significant group with the greatest means (indicated by the left, uppermost bar) shares no populations with the non-significant group having the smallest means (the right, lowermost bar), so differences between those groups are meaningful.

The target symbols and adjacent maps in Figures 4.4-4.11 depict size categories of population means for each character and their distributions. Identical symbols denote populations in the same fifth of the overall range of means. Empty targets are in the lowest fifth of the overall range; completely filled targets are in the highest fifth. In many instances means in different fifths of the range are not statistically different based upon STP as indicated by the vertical bars on the figures. Nevertheless, they indicate that sizes or numbers of some characters are associated with locality and/or species-group.

As indicated by the short vertical lines and targets in Figures 4.4 and 4.5, *Choristoneura occidentalis* (excluding atypical populations C11 and C16) and the hybridizing population of *C. retiniana* (R1) have the longest and broadest wings. Most other populations comprise a few large sets whose means are not significantly different. Populations of *C. spaldingiana*, *C. lambertiana* near *subretiniana*, and *C. c. carnana* are the smallest.

Numbers of basitarsal spines on the hindleg are not significantly different among all populations of *C. occidentalis* (except C11 and C16) and the hybridizing population of *C. retiniana* (R1). They are smallest and not significantly different among all populations of *C. spaldingiana* (Fig. 4.7).

In general, the numbers of mid- and hindleg basitarsal spines are closely related to size (as wing length) except among populations of the "lambertiana" group, where spine numbers are independent of size and approximately equal. In the "lambertiana" group, the correlation coefficient of midleg basitarsal spines and wing length is -0.007, and that of the hindleg spines and wing length is 0.033. In the "carnana" group, these values are 0.492 and 0.556; in the "retiniana" group, they are 0.676 and 0.669.

The uncus tip widths of populations of *Choristoneura l. ponderosana* are the broadest (Fig. 4.10). Those of most of the other populations are not significantly different. Basal widths (Fig. 4.8) tend to be smallest among populations of *C. spaldingiana* and the "lambertiana" group and greatest in the "carnana" group and *C. retiniana* populations.

Differences in uncus constriction widths (Fig. 4.9) are less significant, but the uncus of the "lambertiana" group tends to be the most constricted.

The least variable character is the number of cornuti (Fig. 4.11), where the only significant differences in means are those between populations with the most and the fewest. There is little or no relationship to size (Table 4.1).

Results of discriminant function analysis (d.f.) are presented in Tables 4.3-4.5 and Figures 4.12-4.14. Figures 4.12 and 4.13 show the first and second canonical variables, respectively. These figures illustrate mean discriminant scores of the populations (population centroids), scores of the individuals, the individual variation within the populations, and the overlap of individual scores among different populations. Each centroid is indicated by the symbol (as in Figure 4.1) for the population. The short vertical lines are the individual scores, and the horizontal line indicates the 95% confidence interval about the mean (centroid).

Figure 4.14 is an overall summary of the Euclidean distance between population centroids, based on all seven canonical variables. This figure, a Primnet, connects each population to its nearest neighbor until all are linked to one another (Prim 1957). The lengths of lines connecting adjacent populations are proportional to the Euclidean distance between them.

Tables 4.3 and 4.4 summarize the most probable classifications of individuals by d.f. An individual is "correctly" classified if its most probable classification is to a population with the same *a priori* classification of that moth at a given taxonomic level. An individual may be classified incorrectly at low taxonomic levels (population or subspecies) but correctly classified at higher levels (species or species-group).

Overall, only 38.3% are placed in the correct population, 73.0% are correct at the subspecific level (when *Chortistoneura lambertiana* near *subretiniana* populations are included in *subretiniana*), 75.2% are placed in the correct species, and 82.9% are classified correctly to species-group (Table 4.4). Table 4.5 shows how individuals of the nine *a priori* subspecies are classified by d.f. In Table 4.5, populations L7 and L8 (near *subretiniana*) are included in *C. l. subretiniana* because their centroids cluster with those of *subretiniana* and the percentage of correct subspecific classifications improves from 69.5 to 79.3% when combined. No individuals can be classified as *C. l. lambertiana*, because the only population of *lambertiana* (L3) is represented by a single specimen.

The first three principal components account for 86.5% of the total variation. The first (PRIN1) accounts for 58.7% of the variation and is associated with size. The characters correlated with wing length, an indicator of size, contribute approximately equally to PRIN1, with the less correlated characters, uncus constriction and tip widths, contributing about 75% and 40% as much, respectively. PRIN2, accounting for 16.4% of the total variation, is associated with uncus tip and constriction widths and numbers of basitarsal spines. PRIN3 is associated with the three uncus widths and accounts for 11.4% of the total variation.

Figures 4.15-4.17 are pairwise plots of the centroids of the first three principal components. Figure 4.18 is a "ball and stick" diagram of the first three principal components, and Figure 4.19 is a Primnet summarizing the Euclidean distance between population centroids, based on all six principal components. Although they are not illustrated, the individual variation within and overlapping among populations are similar to those of the d.f. scores.

Scattergrams (Figs. 4.15-4.17) of the centroids of the principal components or the canonical variables of d.f. form clusters of populations generally corresponding to the *a priori* subspecific designations based upon synthetic pheromone attraction, coloration, and wing pattern. Populations failing to cluster with others of the same *a priori* classification are: the hybridizing population of *Choristoneura retiniana* (R1), which clusters with most populations of *C. occidentalis*; the northern Rockies populations of pine-feeding moths (L7 and L8), which form a cluster with the populations of *C. l. subretiniana*; and the atypical populations of *C. occidentalis* (C11 and C16). The nearest neighbor of population C16 is the population of *C. c. carnana* (C6), both of which are remote from the cluster of typical *C. occidentalis* populations. C11 is in the cluster of *C. c. californica* populations. The hybridizing population of *occidentalis* (C1) falls within the cluster of typical *C. occidentalis* populations.

Often, the nearest neighbors of many of the subspecific clusters of populations are subspecific clusters of other species-groups. This indicates that differences in the seven characters among the subspecies within each species-group are sometimes greater than differences from subspecies of other species-groups.

DISCUSSION

The taxonomic confusion among the conifer-feeding *Choristoneura* species arose from several factors. Some of the available names have been misapplied; individual or entire populations of moths of one subspecies may closely resemble those of another; hybridization between species-groups occurs; and geographic and temporal variation is substantial often without obvious distributional patterns. Powell's (1980) review clarified the nomenclatural status, but correct application of the names continues to be a problem.

During preliminary field studies, Powell and I found that these moths are more widely distributed and with greater geographic variation than had been recognized previously. Although often reliable, identifications based upon conspicuous external features were sometimes impossible. It was only with the assistance of the chemical attractants that we were able to associate some populations with others.

Some of the eight morphological characters examined in this study had been used in descriptions; others had not been considered previously. Because of geographic variation and because most measurements are highly correlated with size, I opted to examine a few characters among many populations rather than more characters and fewer populations with more individuals.

None of the eight characters distinguishes moths at any taxonomic level with certainty, and most are extremely variable. In general, however (and with numerous exceptions), the largest moths with the most basitarsal spines on the hind- and midlegs are *Choristoneura occidentalis*; the smallest with the fewest spines are *C. c. carnana* and *C. spaldingiana*.

Choristoneura l. ponderosana has the broadest uncus tip, and moths of all populations in the "lambertiana" group have moderately spinose hind- and midleg basitarsi, usually 12 to 16 spines. The number of cornuti is the least important character examined and is more variable than Freeman's (1967) counts from fewer populations.

With the constituent populations combined, Appendix 4.3 contains descriptive statistics of the most diagnostic characters of the subspecies and statistics of alternative groups using a narrower definition of the subspecies. The size categories (e. g., broad, narrow, long, short) are arbitrary and based on gaps in the range of means, and populations in each category are not statistically different from one another, using *t*-tests and Bonferoni's inequality. Included are some ratios of characters that may be more useful than the individual characters. Other combinations of characters are also diagnostic.

Most moths in the "carnana" group have more basitarsal spines on the hindleg than midleg; most *C. spaldingiana* have more spines on the midleg basitarsus. *Choristoneura l. ponderosana* has the narrowest wings (using wing length to width ratio as an index) and the "carnana" group has the broadest (Table 4.6, Appendix 4.3). (*Choristoneura l. lambertiana* is represented by too few specimens for meaningful comparisons.) Table 4.6 is not sound statistically, in that the ratios were treated as if they were normally distributed, and pairwise *t*-tests were used as the multiple comparison criterion to indicate non-significant groups of subspecies. It is included merely to suggest the differences among the ratios.

The uncus shape is extremely useful for differentiating many groups. In Figure 4.20, population means of maximum uncus tip widths are plotted against minimum constriction widths after adjusting for size by dividing both measurements by maximum basal width. In the entire "carnana" group, *C. retiniana*, and the coastal pine moths, the uncus typically is constricted with a narrow tip. I refer to this as the "normal" uncus shape in Appendix 4B. In the remainder of the "lambertiana" group, it is variably constricted with a moderate to broad tip depending upon subspecies; and it is less constricted with a moderately broad tip in populations of *C. spaldingiana.*

Individual variation is great, so that shape of the uncus of some males more closely resembles the average shape of another species or subspecies than its own species or subspecies. This is true of almost all characters and is reflected in the plots of the individual values of the first two canonical variables of discriminant function analysis (d.f.) (Figs. 4.12 and 4.13), where the individual values of the 40 populations overlap considerably.

A "correct" classification is one in which the *a posteriori* classification by d.f. agrees with the *a priori* classification. When seven of the eight characters are considered together by d.f., only 38.3% of classifications are correct at the population level (Table 4.4). Discrimination of populations is poor because many individuals of a population are very similar to those of related populations. The most probable classification to subspecies is correct for 73.0% of all moths when populations L7 and L8 are included in *Choristoneura l. subretiniana*. The improvement to 75.2% correct specific classifications is small because the differences among subspecies are almost as great as those among the recognized species. This suggests that subspecific names be retained in these moths.

The classifications (Table 4.5) and Primnets based on d.f. and principal components (p.c.) (Figs. 4.14 and 4.19, respectively) reveal most sources of discrepancies between *a priori* and *a posteriori* classifications. Most populations are closely linked to other populations of the same *a priori* subspecific classification, with a few exceptions.

Two atypical *C. occidentalis* populations, C11 in eastern Nevada and C16 near Flagstaff, Arizona, account for 11 of the 23 incorrect classifications among all *C. occidentalis* populations. Most of the misclassified individuals were placed in populations of subspecies of *C. carnana*. Coloration and patterns of many moths in C16 approximate the distinctive "carnana" phenotype. In the Primnets, population C16 links with the single population of *C. c. carnana* (C6).

Two specimens of *Choristoneura* reared from larvae on Douglas-fir from Coahuila, Mexico, and not included in this phenetic study are very similar in phenotype to *C. carnana*. These specimens and population C16 suggest that moths of the *C. carnana* phenotype may have had a southern distribution and persist in a few areas where they are isolated from moths of the more typical *C. occidentalis* phenotype. Population C15, with some moths resembling those of C16 and separated from it by less than 100 km, apparently has more contact with typical *C. occidentalis*. C15 is among the cluster of typical *C. occidentalis* populations. The interrelationships among the Mexican population, populations C15 and C16, *C. carnana*, and typical *C. occidentalis* need further investigation.

The other atypical *C. occidentalis* population (C11) is linked to the cluster of *C. c. californica* populations, but it is closer to the *C. occidentalis* cluster than is C16, the "carnana"-like population. Two of ten C11 individuals are correctly classified to population, two are placed in other *C. occidentalis* populations, and the remainder are variously classified.

Choristoneura c. californica, as used here to include Sierra Nevada populations, is apparently closely related to *C. occidentalis*. Only 37.8% of the moths in populations of *C. c. californica* are correctly classified to subspecies, and most incorrect classifications (12 of 23) are to *C. occidentalis*, its neighbor in the Primnets. The distance between the cluster of *C. c. californica* populations and many of the *C. occidentalis* populations is less than that between the most distant populations in the *C. occidentalis* cluster.

The Sierra Nevada populations of the "carnana" group (C4-5), whose wing pattern and coloration is more similar to to *C. c. carnana* than to *C. c. californica*, show a closer relationship to *californica* than to *carnana* on the basis of the characters considered in this study.

The other major source of incorrect classifications is *C. retiniana* (R1-R3). Only 6 of 21 are classified to subspecies correctly, with most others (10 of 15) classified as *C. occidentalis*.

The presumably hybridizing population of *C. retiniana* (R1) is among the cluster of *C. occidentalis* populations. Seven of nine R1 individuals were classified as *C. occidentalis*. The nearest neighbor of R1 is the sympatric hybridizing population of *C. occidentalis* (C1). (In a separate discriminant function analysis, I used a set of population means to represent the population for classification to species-group. The only misclassified population was that of the hybridizing population of *C. retiniana* (R1), which was classified in the "carnana" group.)

Volney *et al.* (1983) and Liebhold (1986) determined that hybrids of *Choristoneura occidentalis* and *C. retiniana* often have the brown larval coloration of the former species, and that the green coloration of *C. retiniana* larva increases by backcrossing to *C. retiniana.* Similarly, apparently there is simple dominance of inheritance of the *C. occidentalis* sex pheromone over the *C. retiniana* sex pheromone in F_1 hybrid females (Sanders *et al.* 1977, Liebhold *et al.* 1984, De Benedictis *et al.* Part 5 this vol.). Assuming that some or most specimens of population R1 resulted from *C. retiniana* interbreeding with *C. occidentalis,* the analyses here suggest that many morphological characteristics of adult *C. occidentalis* are also dominant.

For all characters, the means of C1 and R1, although not significantly different, differ in the same direction as other sympatric members of their species-groups. R1 is smaller and with fewer basitarsal spines than C1, for example.

The other populations of *Choristoneura retiniana* are linked in the Primnets by R2 with the cluster of *C. c. californica* populations. The distance between R2 and the *C. c. californica* cluster is shorter than that between R2 and R3. This would suggest that R2 interbreeds with *C. c. californica*, but pheromone trapping near the collection sites in 1981 indicated that R3 was sympatric with *C. c. californica* and R2 was not. Thus, the apparent relationship of R2 to *C. c. californica* is either coincidental or due to hisorical contact and hybridization with *C. c. californica.* Alternatively, the Primenet distance and infered relationship among these populations might have changed had I included a *C. c. californica* sample from the vicinity of the R3 collection sites in this study.

Overall, the three populations of *C. retiniana* and two atypical populations of *C. occidentalis* comprise just 9.6% of all individuals but account for 27.5% of the incorrect subspecific classifications.

Assuming that distances are associated with relationships among moths, rather than the coincidental results of random variation, d.f. and p.c. based on the seven morphological characters indicate that, in general, the constituent populations of the *a priori* groups are related.

In the "carnana" group, expanding the concept of *C. carnana* to include population C16 could be justified on the basis of multivariate analyses and their general appearances, but biological evidence is needed to confirm this association.

Choristoneura c. californica, which more closely resembles *C. carnana* than *C. occidentalis* by facies, is closer to the latter on the basis of the morphological characters. This suggests that if subspecific status of *californica* is warranted, *C. occidentalis* also should be considered a subspecies of *C. carnana*. On the other hand, maintaining the specific status of *C. occidentalis* is indicated by size and phenotypic polymorphism in population C15, in which some individuals resemble either species, and by the distance between the clusters of typical *C. occidentalis* and *C. c. californica.* The Primnets and "ball and stick" diagram (Fig. 4.18) illustrate that the distance between the two clusters is comparable to those between some subspecific clusters of two different species-groups.

The relationship of the small mottled moths of the "carnana" group, represented by population C11 from the Schell Creek Range, Nevada, to the remainder of the "carnana" group needs clarification. It may be a sibling species or temporal variant of more typical *C. occidentalis*. Failure of C11 to cluster with other *C. occidentalis* populations emphasizes its uncertain status.

In the "lambertiana" group, d.f. and p.c. indicate that the concept of *Choristoneura l. subretiniana* should be broadened to include northern Rocky Mountains populations, as represented by L7 and L8. The first few canonical variables and principal components of *C. l. subretiniana* (*sens. lat.*) are close to the single specimen of *C. l. lambertiana* (L3) (Figs. 4.12-13 and 4.15-18), but additional specimens of *C. l. lambertiana* are needed to assure that this is not coincidental. The retention of *subretiniana* as a subspecies distinct from *lambertiana* can be supported by differences of larval hosts and wing pattern and coloration.

The distance between these populations and the cluster of *C. l. ponderosana* populations is greater than most distances between subspecific clusters of different "species" groups. However, some of the distances between populations within the *C. l. ponderosana* cluster are also greater than many interspecific distances. These distances might be narrowed if additional populations were examined. In the cluster of typical *C. occidentalis*, for example, distances between the least similar populations are comparable, but intermediate populations make the distances less apparent.

Populations of *C. l. ponderosana* from east of the continental divide, including the type series, were outside the region considered in this study, and other populations were represented by too few individuals. Their inclusion might shorten links in the cluster. Similarly, including a population sample from Wyoming, where specimens appearing intermediate between the phenotypes of *C. l. ponderosana* and northern Rocky Mountains pine moth occur, may shorten the distance between the *C. l. ponderosana* and *subretiniana* clusters.

Assignment of the coastal Oregon and Washington pine moth (L1 and L2) to the "lambertiana" group is tentative. The multivariate analyses (d.f. and p.c.) underscore that their relationship to the remainder of the group needs clarification. Gray and Gries (1993) characterized the pheromone of a presumedly undescribed *Choristoneura* on Lodgepole Pine in British Columbia, which may be the same as populations L1 and L2. Chemical attraction of L1 and L2 has not been tested.

The difference between populations R1-R3 of *C. retiniana* and the remainder of the "retiniana" group is sufficient to warrant retention of the name *C. spaldingiana*, at least as a subspecies. Obraztsov, who described *spaldingiana* (Obraztsov 1962), affixed a label bearing the name *C. spaldingiana* to a specimen in the American Museum of Natural History from Lake Tahoe, California, so his concept of the "species" encompassed a wide geographic range similar to that used here (Fig. 4.1).

Although the results suggest retention of the names of taxa recognized by Powell (1980), they provide little information about their taxonomic levels. Because chemical attraction, larval coloration, and the genus of preferred larval host separate the western taxa into three groups, a conservative treatment would recognize the three species-groups as species until additional evidence indicates that splitting the groups into more species is warranted.

With the possible exception of moths from the regions where hybridization is recent and frequent, d.f. is an efficient means of identifying subspecies of *Choristoneura*. Over 75% of the moths were correctly classified into nine subspecies. The proportion of unknown moths correctly classified likely would be lower, but by using more characters, the proportion of correct classifications would likely increase. Daly and Balling (1978), for example, using 25 characters and 10 or 20 individuals achieved 100% discrimination between hives of Africanized and European honeybees. With both honeybees and spruce budworms, 100% discrimination of individuals is unlikely because of random variation and the effects of hybridization.

Promising characters not examined in this study are the length of the aedeagus tip, which Dang (1985) found to differ among species; dimensions of other genitalic features; and lengths and widths of legs, abdominal sclerites, and other external body parts. Female genitalia also may be useful for identifying taxa.

Although reliable identification of conifer-feeding *Choristoneura* species and subspecies by d.f. would require considerable effort to expand the data set to include more populations and characters, the differences between states of many of the characters examined here are sufficient to differentiate moths of the three species-groups at most locations. Thus, simply determining differences of the most diagnostic characters (*e.g.*, basitarsal spines and uncus shape) at other localities may provide enough information to identify moths accurately if their collection site is known.

LITERATURE CITED

Conover, W.J., M.E. Johnson, and M.M. Johnson. 1981. A comparative study of tests for homogeneity of variances, with applications to the outer continental shelf bidding data. Technometrics 23:351-361.

Daly, H.V., and S.S. Balling. 1978. Identification of Africanized honeybees in the western hemisphere by discriminant analysis. J. Kansas Entomol. Soc. 51:857-869.

Dang, P.T. 1985. Key to adult males of conifer-feeding species of *Choristoneura* Lederer (Lepidoptera: Tortricidae) in Canada and Alaska. Canad. Entomol. 117:1-5.

Daterman, G.E., L.L. Sower, R.E. Stevens, and D.G. Fellin. 1995. Pheromone chemistry and response behavior of conifer-feeding *Choristoneura* in the western United States. Part 3 this volume.

De Benedictis, J.A., A.M. Liebhold, and J.A. Powell. 1995. Studies of pheromone attraction and inheritance using virgin females of *Choristoneura* species, in the central Sierra Nevada, California. Part 5 this volume.

Freeman, T.N. 1967. On coniferophagous species of *Choristoneura* (Lepidoptera: Tortricidae) in North America. I. Some new forms of *Choristoneura* allied to *C. fumiferana*. Canad. Entomol. 99:449-455.

Gabriel, K.R. 1964. A procedure for testing the homogeneity of all sets of means in analysis of variance. Biometrics 20:459-477.

Gray, T.C., and G. Gries. 1993. Sex pheromone components of an undescribed *Choristoneura* species on lodgepole pine in British Columbia. J. Entomol. Soc. Brit. Col. 90:13-18.

Hodges, R.W., T. Dominick, D.R. Davis, D.C. Ferguson, J.G. Franclemont, E.G. Munroe, and J.A. Powell (eds.). 1983. Check list of the Lepidoptera of America north of Mexico. E.W. Classey, Ltd., London. xxiv + 161 pp.

Liebhold, A.M. 1986. Larval coloration of hybrids between *Choristoneura occidentalis* and *C. retiniana* (Lepidoptera: Tortricidae). Canad. Entomol. 118:857-860.

Liebhold, A.M., W.J.A. Volney, and W.E. Waters. 1984. Evaluation of cross-attraction between sympatric *Choristoneura occidentalis* and *C. retiniana* (Lepidoptera: Tortricidae) populations in south-central Oregon. Canad. Entomol. 116:827-840.

Miller, W.E. 1977. Wing measure as a size index in Lepidoptera: the family Olethreutidae. Ann. Entomol. Soc. Am. 70:253-256.

Obraztsov, N.S. 1962. New species and subspecies of North American Archipini, with notes on other species (Lepidoptera, Tortricidae). Am. Mus. Novit. 2101:1-26.

Powell, J.A. 1980. Nomenclature of Nearctic conifer-feeding *Choristoneura* (Lepidoptera: Tortricidae): Historical review and present status. USDA For. Serv. Pacific Northwest Forest and Range Experiment Station General Technical Report PNW-100; 18 pp.

Powell, J.A. and J.A. De Benedictis. 1995a. Biological Relationships: Host tree preferences and isolation by pheromones among allopatric amd sympatric populations of western *Choristoneura*. Part 2 this volume.

Powell, J.A. and J.A. De Benedictis. 1995b. Evolutionary interpretation, taxonomy, and nomenclature. Part 7 this volume.

Power, D.M. 1970. Geographic variation of red-winged blackbirds in central North America. Univ. Kansas Mus. Nat. Hist. Misc. Publ. 19:1-83.

Prim, R.C. 1957. Shortest connection networks and some generalizations. Bell System Technical Journal 36:1389-1401.

Sanders, C.J., G.E. Daterman, R.F. Shepherd, and H. Cerezke. 1974. Sex attractants for two species of western spruce budworms, *Choristoneura biennis* and *C. viridis* (Lepidoptera: Tortricidae). Canad. Entomol. 106:157-159.

Sanders, C.J., G.E. Daterman, and T.J. Ennis. 1977. Sex pheromone responses of *Choristoneura* spp. and their hybrids (Lepidoptera: Tortricidae). Canad. Entomol. 109:1203-1220.

Volney, W.J.A., A.M. Liebhold, and W.E. Waters. 1983. Effects of temperature, sex, and genetic background on coloration of *Choristoneura* spp. (Lepidoptera: Tortricidae) populations in south-central Oregon. Canad. Entomol. 115:1583-1596.

Volney, W.J.A., A.M. Liebhold and W.E. Waters. 1984. Host associations, phenotypic variation, and mating compatibility of *Choristoneura occidentalis* and *C. retiniana* (Lepidoptera: Tortricidae) populations in south-central Oregon. Canad. Entomol. 116:813-826.

Table 4.1. Correlation And Covariance Matrix Of Eight Morphometric Characters

No.	Character	No.	Character
1.	Wing Length	5.	Maximum Width of Uncus Base
2.	Wing Width	6.	Minimum Uncus Width
3.	Number of Midleg Basitarsal Spines	7.	Maximum Width of Uncus Tip
4.	Number of Hindleg Basitarsal Spines	8.	Number of Cornuti

R / Cov

	1.	2.	3.	4.	5.	6.	7.
2.	0.921 34.400						
3.	0.599 20.485	0.642 11.058					
4.	0.622 23.238	0.663 12.471	0.824 14.161				
5	0.689 5.176	0.747 2.832	0.556 1.908	0.564 2.078			
6.	0.456 2.278	0.473 1.192	0.274 0.624	0.223 0.544	0.638 0.311		
7.	0.321 2.698	0.270 1.145	0.081[a] 0.309	0.112 0.461	0.169 0.138	0.272 0.148	
8.	0.154 1.332	0.114[b] 0.491	0.083[a] 0.333	0.099 0.423	0.160 0.139	0.162 0.094	0.041[a] 0.039

[a] Correlation not significantly greater than zero at $\alpha = 0.05$.

[b] Correlation not significantly greater than zero at α =.0.01.

Table 4.2. Tests For Equality Of Means By Anova And Variances By Lev1:Med

Character	Wing Length	Wing Width	Midleg Basitarsal Spines
ANOVA			
F-ratio	16.01	17.82	12.72
d.f.	38, 344	38, 344	38, 340
p-value	nil	nil	nil
Lev1:Med			
F-ratio	4.79	5.22	2.38
d.f.	38, 344	38, 344	38, 340
p-value	$8.67X10^{-16}$	$1.19X10^{-17}$	$2.21X10^{-5}$

Character	Hindleg Basitarsal Spines	Max. Uncus Basal Width	Minimum Uncus Width
ANOVA			
F-ratio	19.37	12.20	5.46
d.f.	38, 338	38, 289	38, 289
p-value	nil	nil	$7.20X10^{-18}$
Lev1:Med			
F-Ratio	4.01	2.10	2.19
d.f.	38, 338	38, 289	38, 289
p-value	$2.39X10^{-12}$	$3.47X10^{-4}$	$1.62X10^{-4}$

Character	Max. Uncus Tip Width	Cornuti
ANOVA		
F-ratio	4.94	2.04
d.f.	38, 289	38, 260
p-value	$8.89X10^{-16}$	$6.22X10^{-4}$
Lev1:Med		
F-ratio	3.18	2.79
d.f.	38, 289	38, 260
p-value	$1.78X10^{-8}$	$9.02X10^{-7}$

Table 4.3. Correct Classifications By Discriminant Function Analysis

	N	% of Total	Taxa Recognized[1]
"Carnana" group:	142		
to population:	49	34.5	
to subspecies:	91	64.1	carn, calif, occ
to species:	91	64.1	carn, occ
to species-group:	113	79.6	
"Lambertiana" group:	82		
to population:	40	48.8	
to subspecies:	57	69.5	coast, lamb, sub, nr. sub, pond
to subspecies:	65	79.3	coast, lamb, sub, pond
to species:	70	85.4	lamb, coast
to species-group:	73	89.0	
"Retiniana" group:	98		
to population:	36	36.7	
to species:	79	80.6	ret, spald
to species-group:	81	82.7	

[1] See footnote to Table 4.4 for abbreviations of subspecies.

Table 4.4. Correct Classifications By Discriminant Function Analysis

	Taxa N	% of Total	Recognized[1]
All Moths:	322		
to population:	125	38.3	
to subspecies:	235	73.0	occ, carn, calif, coast, lamb, sub, pond, ret, spald
to species	242	75.2	occ, carn, ret (incl. spald.), coast, lamb
to species-group:	267	82.9	

[1]Abbreviations of subspecies:
occ = *C. occidentalis* (incl. hybridizing population);
carn = *C. carnana carnana*;
calif = *C. carnana californica*;
coast = coastal pine-feeding populations;
lamb = *C. lambertiana lambertiana*;
sub = *C. l. subretiniana* (incl. nr. *subretiniana* [nr. sub.] except where indicated);
pond = *C. l. ponderosana*;
ret = *C. retiniana* (incl. hybridizing population);
spald = *C. l. spaldingiana*.

Abbreviations of species: Each is abbreviated by the designation for the nominotypic subspecies above (occ, carn, lamb, ret) and includes all its subspecies.

Table 4.5. Classification Of Individuals By Discriminant Function Analysis

A priori Designation (N)	Classified as (n and % of total):							
	occ	carn	calif	sub	pond	coast	ret	spald
Choristoneura occidentalis (95)	72	4	5	3	0	6	3	2
	75.8	4.2	5.3	3.2	0.0	6.3	3.2	2.1
C. carnana carnana (10)	1	5	0	1	1	2	0	0
	10.0	50.0	0.0	10.0	10.0	20.0	0.0	0.0
*C. carnana californica** (37)	12	0	14	6	1	2	0	2
	32.4	0.0	37.8	16.2	2.7	5.4	0.0	5.4
C. lambertiana lambertiana (1)	0	0	0	0	0	0	1	0
	0.0	0.0	0.0	0.0	0.0	0.0	100.0	0.0
*C. lambertiana subretiniana*** (38)	1	0	0	28	3	2	1	3
	2.6	0.0	0.0	73.7	7.9	5.3	2.6	7.9
C. lambertiana ponderosana (32)	0	0	0	2	30	0	0	0
	0.0	0.0	0.0	6.3	93.8	0.0	0.0	0.0
coastal pine populations (11)	0	0	1	1	0	7	0	2
	0.0	0.0	9.1	9.1	0.0	63.6	0.0	18.2
C. retiniana (21)	10	2	0	0	0	1	6	2
	47.6	9.5	0.0	0.0	0.0	4.8	28.6	9.5
C. spaldingiana (77)	0	1	0	1	2	0	0	73
	0.0	1.3	0.0	1.3	2.6	0.0	0.0	94.8
All moths (322)	96	12	20	42	37	20	11	84
	29.8	3.7	6.2	13.0	11.5	6.2	3.4	26.1

*Includes Sierra Nevada populations of *C. carnana*.

**Includes populations of *C. l. subretiniana* and *C. l.* nr. *subretiniana*.

Table 4.6. Wing Length To Width Ratios By Subspecies[1] (Mean/Standard Deviation)

C. l. ponderosana[a]	*C. spaldingiana*[a,b]	coastal pine[b,c]
2.556 ±0.121	2.519 ±0.098	2.482 ±0.099
C. l. subretiniana[c,d]	*C. retiniana*[c]	*C. c. carnana*[c,d]
2.465 ±0.106	2.462 ±0.114	2.435 ±0.098
C. occidentalis[d]	*C. c. californica*[d]	*C. l. lambertiana*[d]
2.393 ±0.095	2.376 ±0.113	2.269 ±0.186

[1] Wing length to wing width ratios by subspecies. Means and standard deviations of wing length divided by wing width are listed below the names of the nine subspecies. Superscripts indicate groups of subspecies with equal ratios using pairwise *t*-tests at $\alpha = 0.05$.

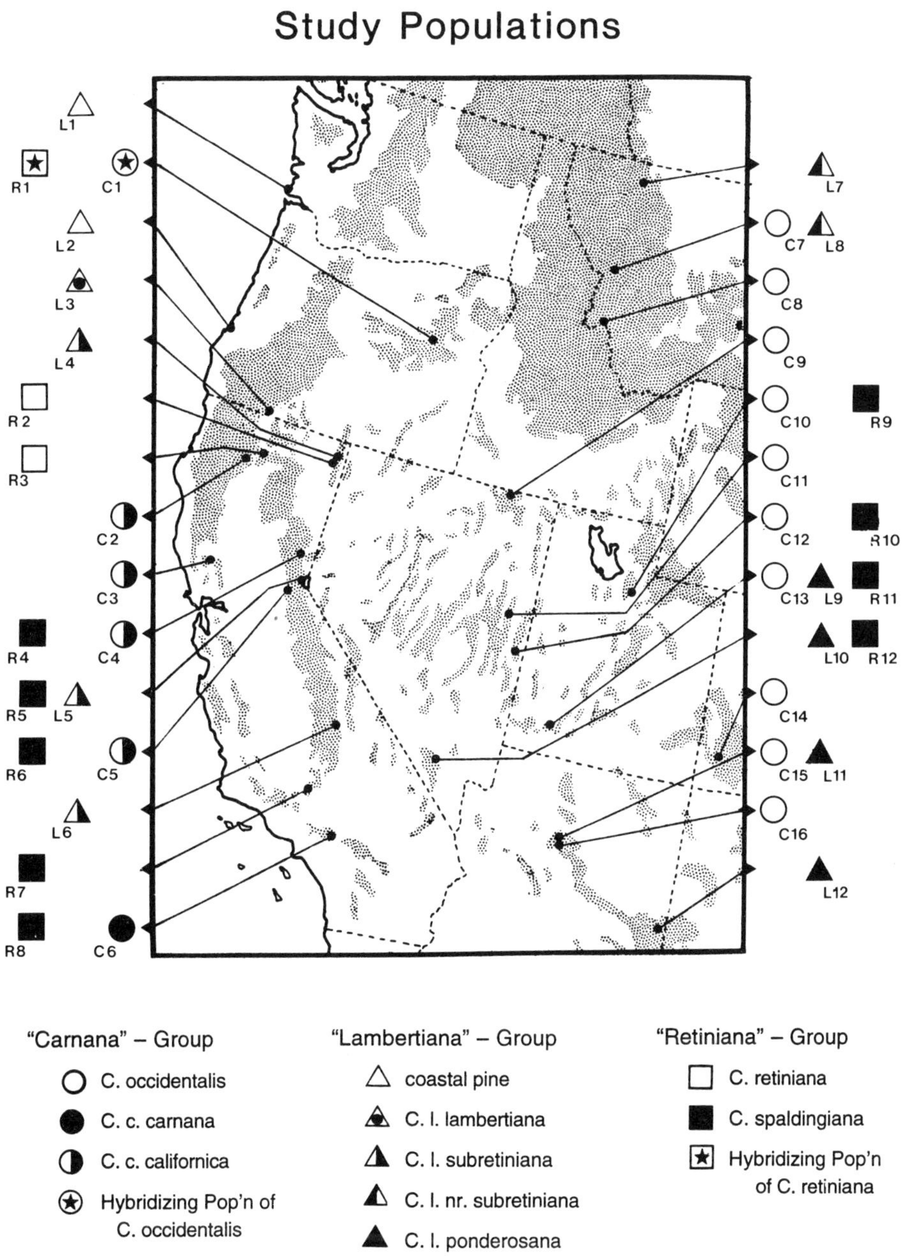

Fig. 4.1. Localities and *a priori* classifications of 40 study populations. Species-groups are arranged in columns, with the "retiniana" group in the outermost and the "carnana" group in the innermost column. Sympatric populations are in rows. On each side of the map, lines connect collection localities to the populations in each row. Shaded area approximates the distribution of conifers.

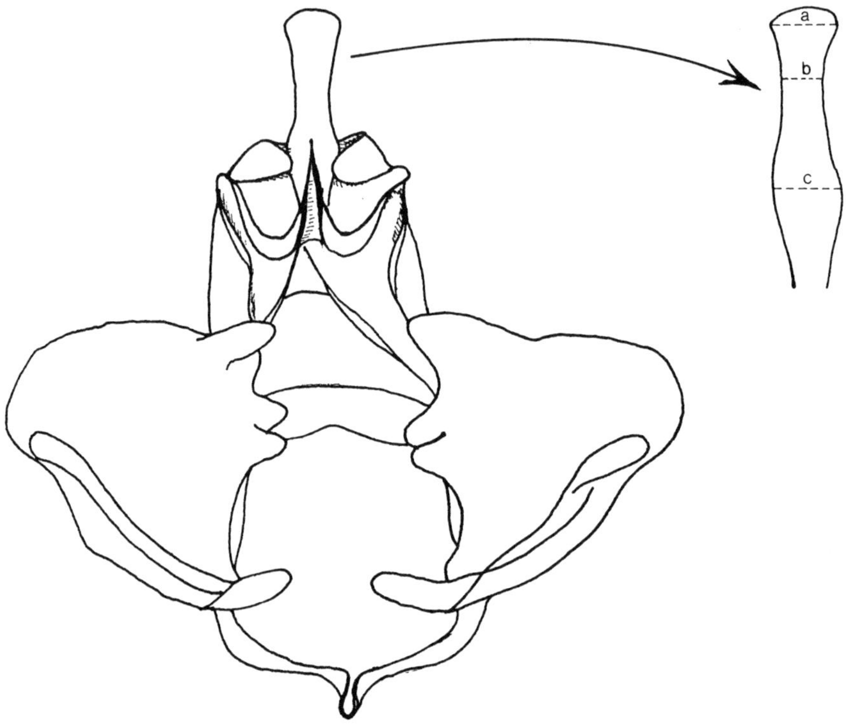

Fig. 4.2. Uncus measurements. The genitalia of *C. occidentalis* with an enlargement of the flattened uncus to the right. The maximum apical width (a), minimum width of constriction (b), and maximum basal width (c) of the projected image of the uncus were measured to 0.1 mm. Note that the uncus is somewhat asymmetrical and that width maxima were measured parallel to the minimum width.

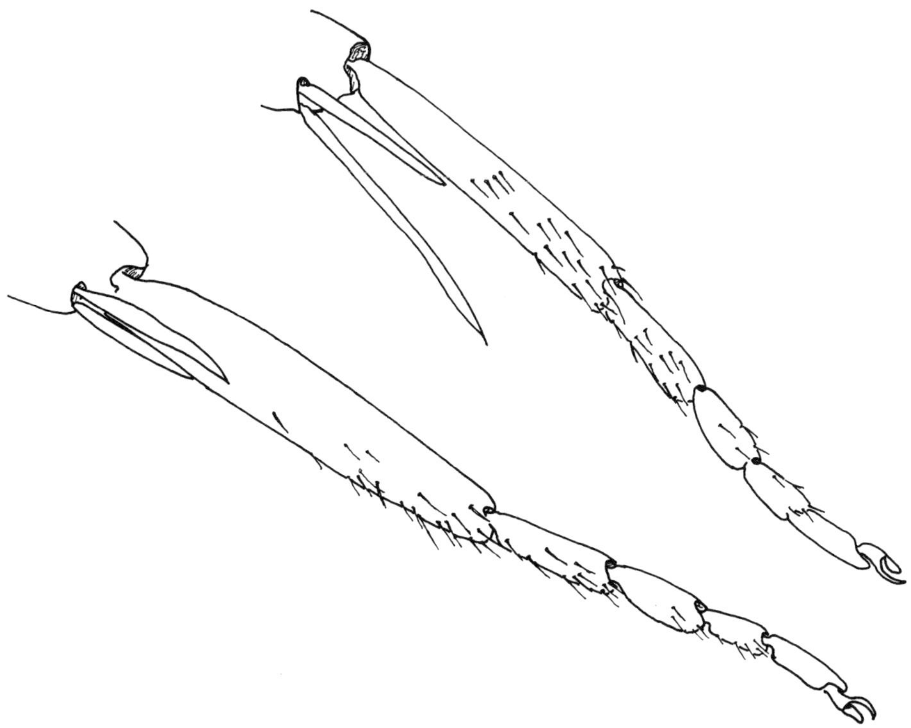

Fig. 4.3 The distribution of tarsal spines on the midleg (above, right) and hindleg of a specimen of *C. occidentalis*. Only counts of the basitarsal spines were used.

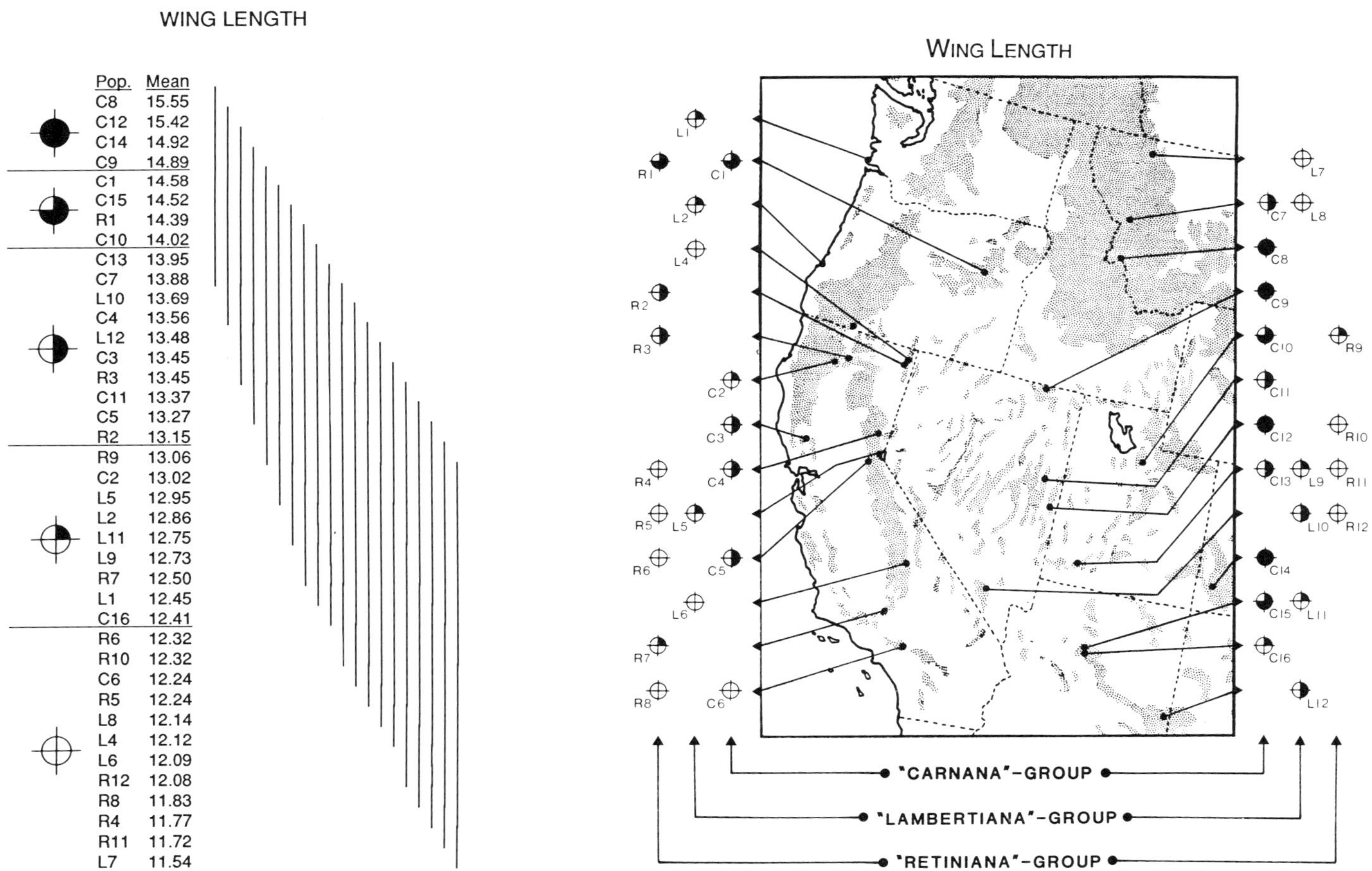

Fig. 4.4. Forewing length. Vertical lines indicate sets of populations with means that are not significantly different at $\alpha = 0.05$ by the simultaneous testing procedure (STP). Maps depict distributions of populations in each fifth of the overall range of means. Empty targets are in the lowest fifth of the range; one filled quadrant indicates populations in the next lowest fifth, etc.

WING WIDTH

Pop.	Mean
C8	6.55
C12	6.55
C1	6.20
C9	6.18
C15	6.17
C14	6.14
R1	5.98
C7	5.89
C4	5.78
C10	5.78
C13	5.78
C3	5.67
C5	5.61
L12	5.57
C11	5.52
R3	5.50
C2	5.42
L10	5.36
L5	5.23
L2	5.23
C16	5.20
R2	5.20
R6	5.10
R9	5.10
C6	5.02
L8	5.01
L1	4.99
R7	4.92
L9	4.92
R5	4.90
L6	4.89
R10	4.86
L4	4.83
L7	4.83
R12	4.82
L11	4.82
R4	4.70
R8	4.69
R11	4.55

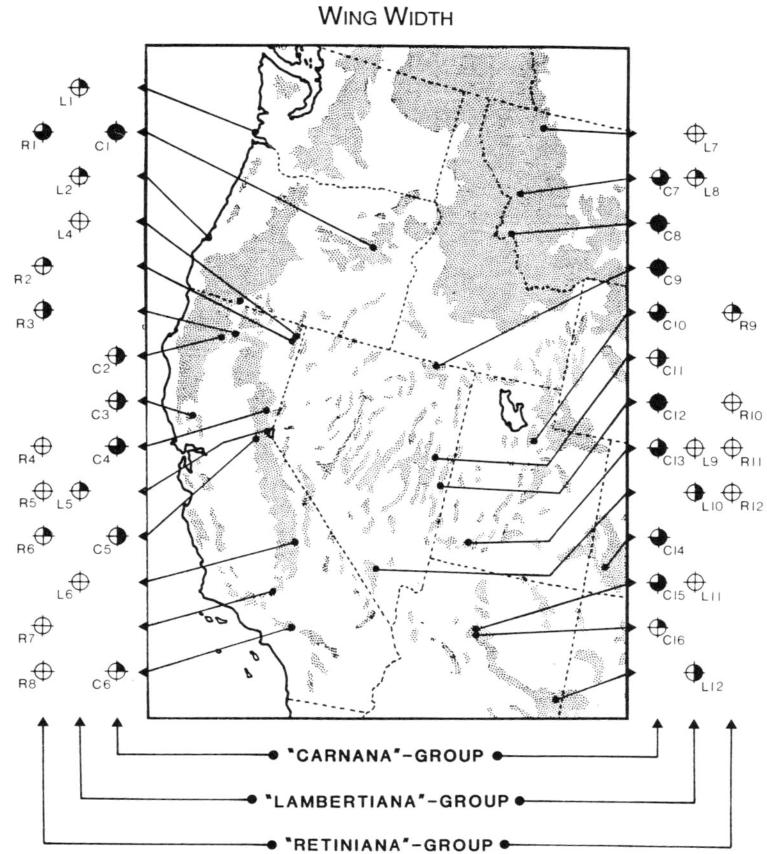

Fig. 4.5. Forewing width. Vertical lines indicate sets of populations with means that are not significantly different at $\alpha = 0.05$ by the simultaneous testing procedure (STP). Maps depict distributions of populations in each fifth of the overall range of means. Empty targets are in the lowest fifth of the range; one filled quadrant indicates populations in the next lowest fifth, etc.

MIDLEG BASITARSAL SPINES

Pop.	Mean
C8	20.20
C12	20.12
C10	20.00
C1	18.70
C9	18.30
C7	18.14
R1	17.70
C13	17.50
C14	17.10
C3	16.50
R3	16.20
C2	16.00
L2	15.80
C4	15.60
C5	15.60
C15	15.40
L1	14.60
L7	14.29
L5	14.04
L8	14.00
C16	13.60
C11	13.50
L6	13.30
R2	13.30
L9	13.10
L12	13.00
L10	12.80
L4	12.40
C6	12.25
L11	12.20
R6	11.90
R9	11.70
R11	10.62
R4	10.50
R5	10.30
R10	10.30
R7	9.95
R8	9.90
R12	9.30

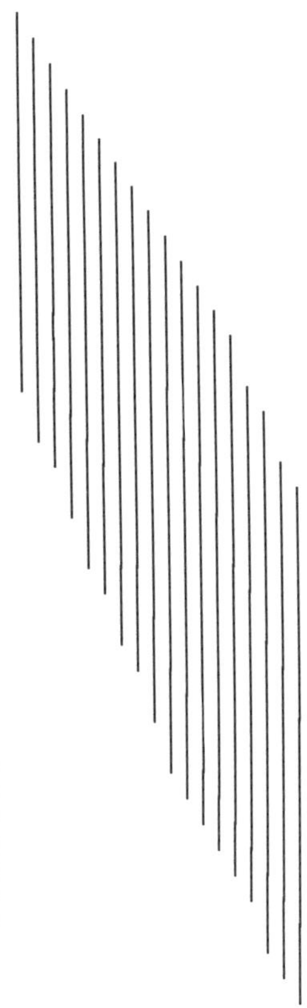

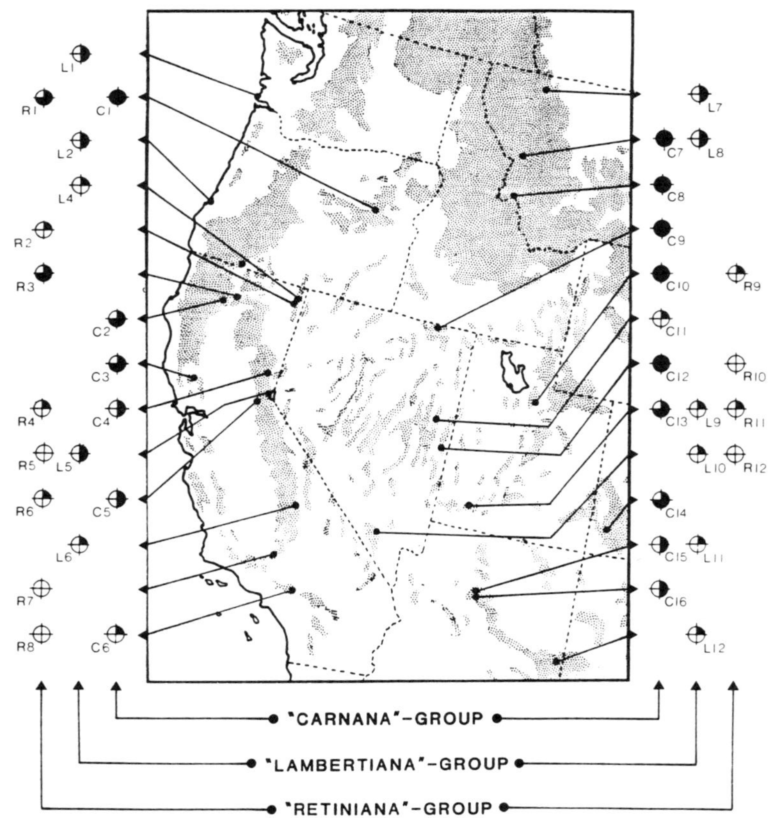

Fig. 4.6. Number of midleg basitarsal spines. Vertical lines indicate sets of populations with means that are not significantly different at $\alpha = 0.05$ by the simultaneous testing procedure (STP). Maps depict distributions of populations in each fifth of the overall range of means. Empty targets are in the lowest fifth of the range; one filled quadrant indicates populations in the next lowest fifth, etc.

HINDLEG BASITARSAL SPINES

Pop.	Mean
C8	21.70
C12	19.88
C9	19.60
C1	19.30
C10	19.10
C7	18.57
R1	17.90
C13	17.80
L2	16.80
C3	16.50
C14	16.50
C2	16.40
C15	16.40
L1	15.50
C5	15.45
C11	15.10
R3	14.70
C4	14.70
L8	14.57
L5	14.50
C16	14.40
L10	14.30
L7	14.00
L6	13.70
L4	13.60
L9	13.30
L11	12.90
L12	12.75
C6	12.45
R2	12.10
R6	11.40
R4	9.80
R9	9.60
R5	9.40
R7	9.30
R10	9.00
R8	8.50
R12	7.70
R11	7.50

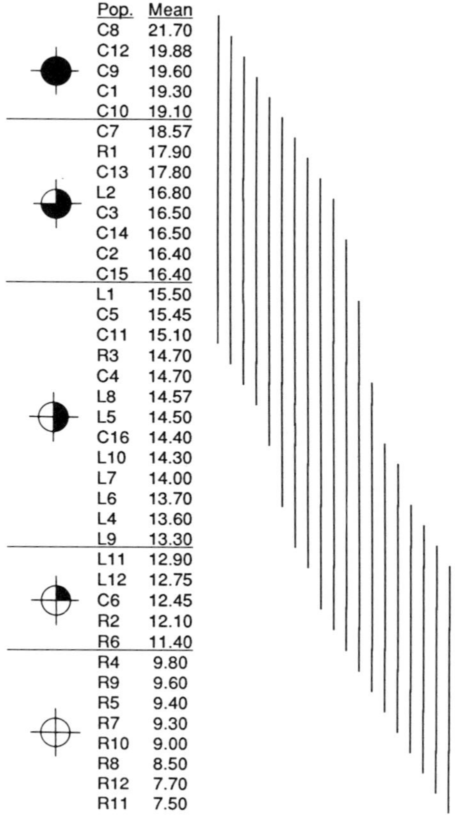

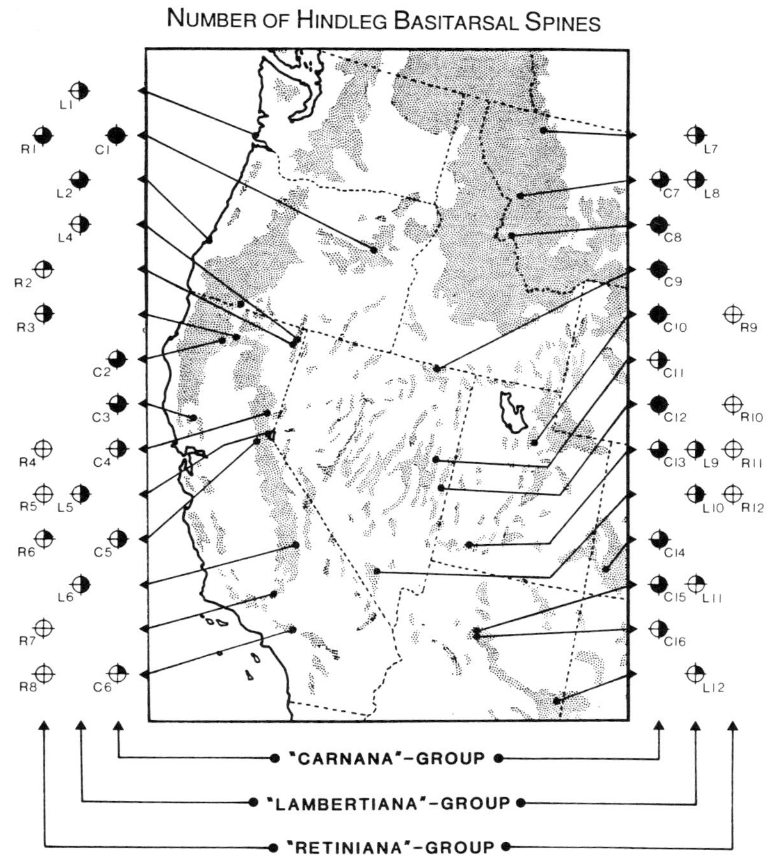

Fig. 4.7. Number of hindleg basitarsal spines. Vertical lines indicate sets of populations with means that are not significantly different at $\alpha = 0.05$ by the simultaneous testing procedure (STP). Maps depict distributions of populations in each fifth of the overall range of means. Empty targets are in the lowest fifth of the range; one filled quadrant indicates populations in the next lowest fifth, etc.

MAXIMUM WIDTH OF UNCUS BASE

Pop.	Mean
C12	7.56
C9	7.33
C8	7.19
R2	7.12
C7	6.99
C13	6.94
C1	6.94
R1	6.92
C15	6.85
R3	6.76
C3	6.66
C14	6.65
C4	6.64
C10	6.55
C5	6.54
C11	6.43
C2	6.31
L2	6.23
C16	6.19
C6	6.16
L1	6.08
L5	6.04
R9	5.84
L10	5.80
R12	5.79
R11	5.71
R7	5.69
R6	5.68
L4	5.68
R8	5.63
R4	5.61
R10	5.44
R5	5.40
L11	5.33
L12	5.26
L8	5.24
L9	5.20
L7	5.18
L6	5.13

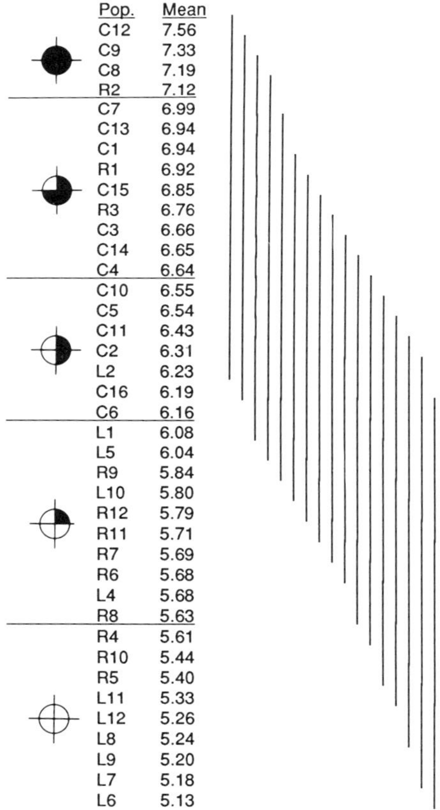

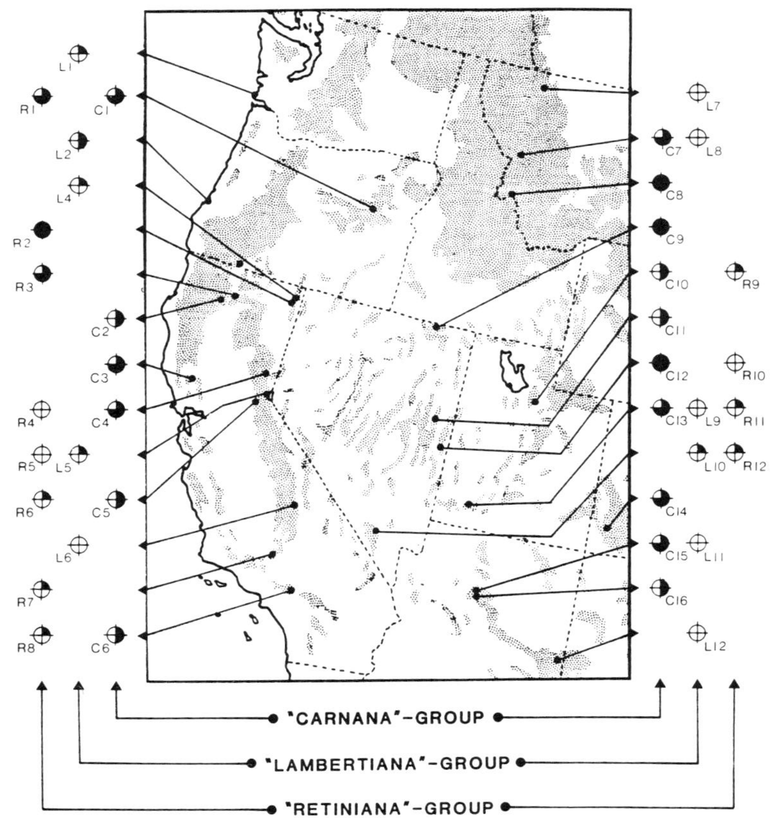

Fig. 4.8. Maximum width of uncus base. Vertical lines indicate sets of populations with means that are not significantly different at $\alpha = 0.05$ by the simultaneous testing procedure (STP). Maps depict distributions of populations in each fifth of the overall range of means. Empty targets are in the lowest fifth of the range; one filled quadrant indicates populations in the next lowest fifth, etc.

MINIMUM WIDTH OF UNCUS CONSTRICTION

Pop.	Mean
C9	5.45
C12	5.44
C8	5.35
C7	5.11
C4	5.06
R3	5.01
R12	4.99
R2	4.98
C1	4.97
C3	4.90
C5	4.89
R1	4.86
R8	4.86
C11	4.85
C16	4.83
C14	4.82
R6	4.81
C15	4.79
R9	4.79
C13	4.77
R10	4.73
R7	4.72
R11	4.70
C10	4.59
L12	4.58
R5	4.55
C6	4.54
L2	4.53
C2	4.50
L1	4.48
L11	4.47
R4	4.44
L10	4.43
L9	4.16
L5	4.16
L4	4.13
L7	3.98
L8	3.98
L6	3.83

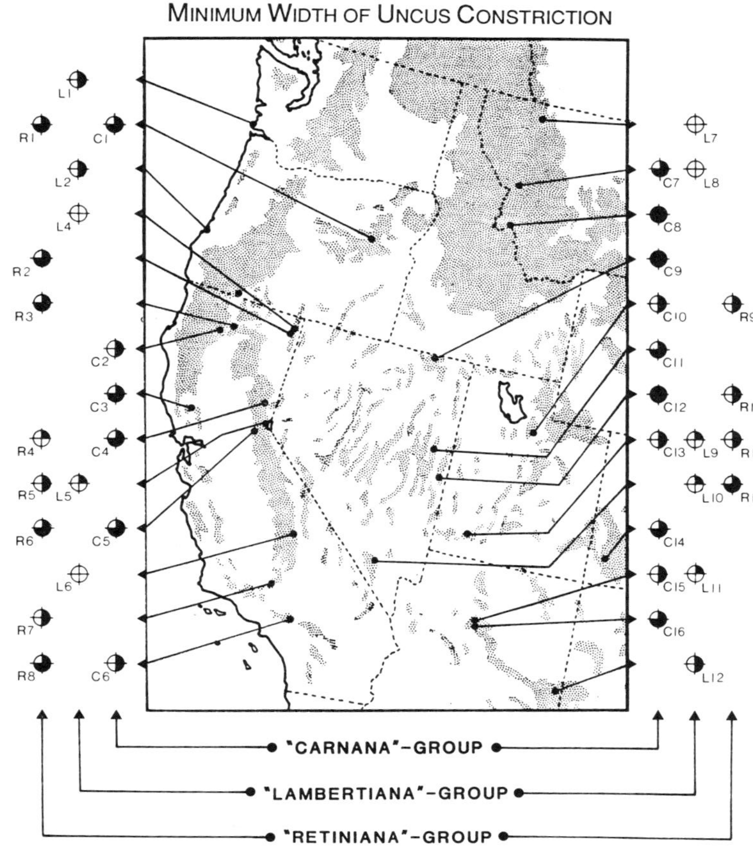

Fig. 4.9. Minimum width of uncus constriction. Vertical lines indicate sets of populations with means that are not significantly different at $\alpha = 0.05$ by the simultaneous testing procedure (STP). Maps depict distributions of populations in each fifth of the overall range of means. Empty targets are in the lowest fifth of the range; one filled quadrant indicates populations in the next lowest fifth, etc.

WIDTH OF UNCUS TIP

Pop.	Mean
L10	9.16
L12	8.96
L11	8.50
L9	8.46
C12	8.08
R9	8.05
R3	7.89
C8	7.88
R10	7.85
R2	7.78
C9	7.67
C7	7.64
C3	7.62
L5	7.61
R8	7.44
C15	7.42
C2	7.39
C1	7.37
R11	7.35
R1	7.31
R6	7.30
L4	7.30
C11	7.29
C5	7.24
R7	7.19
C13	7.18
C14	7.15
R12	7.14
C4	7.10
L6	7.03
L2	6.98
R5	6.97
L8	6.94
C16	6.80
L7	6.74
C6	6.66
R4	6.65
C10	6.55
L1	6.10

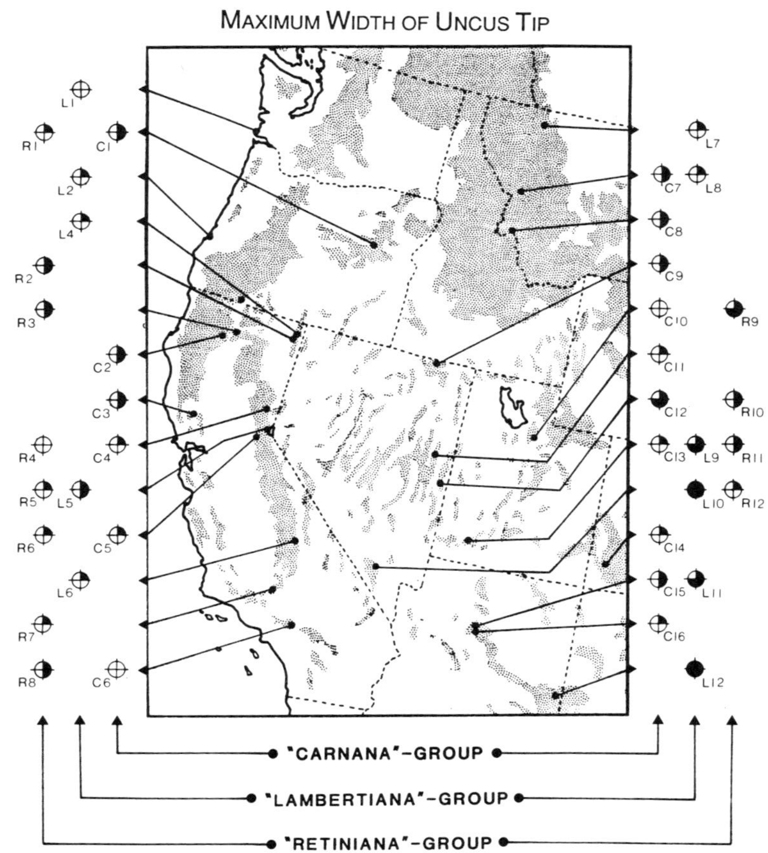

Fig. 4.10. Width of uncus tip. Vertical lines indicate sets of populations with means that are not significantly different at α = 0.05 by the simultaneous testing procedure (STP). Maps depict distributions of populations in each fifth of the overall range of means. Empty targets are in the lowest fifth of the range; one filled quadrant indicates populations in the next lowest fifth, etc.

NUMBER OF CORNUTI

Pop.	Mean
R2	4.63
R10	4.33
C9	4.25
R8	4.14
R7	4.00
C8	4.00
C1	4.00
C12	3.88
C6	3.80
C7	3.71
C2	3.67
C4	3.67
L1	3.67
C5	3.62
R6	3.62
R9	3.63
L4	3.60
R5	3.57
L11	3.50
R12	3.43
L5	3.40
C3	3.38
C13	3.33
R11	3.33
L10	3.29
R3	3.25
C11	3.22
L8	3.17
L2	3.14
C10	3.12
C14	3.11
C15	3.00
L7	3.00
R1	3.00
C16	2.89
L9	2.75
R4	2.75
L6	2.50
L12	2.50

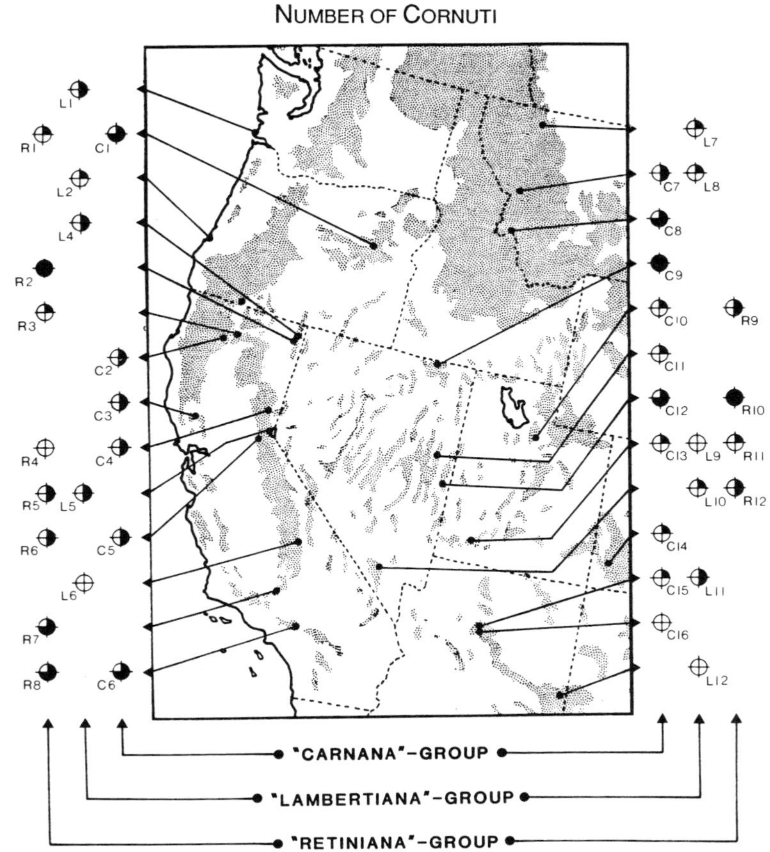

Fig. 4.11. Number of cornuti. Vertical lines indicate sets of populations with means that are not significantly different at $\alpha = 0.05$ by the simultaneous testing procedure (STP). Maps depict distributions of populations in each fifth of the overall range of means. Empty targets are in the lowest fifth of the range; one filled quadrant indicates populations in the next lowest fifth, etc.

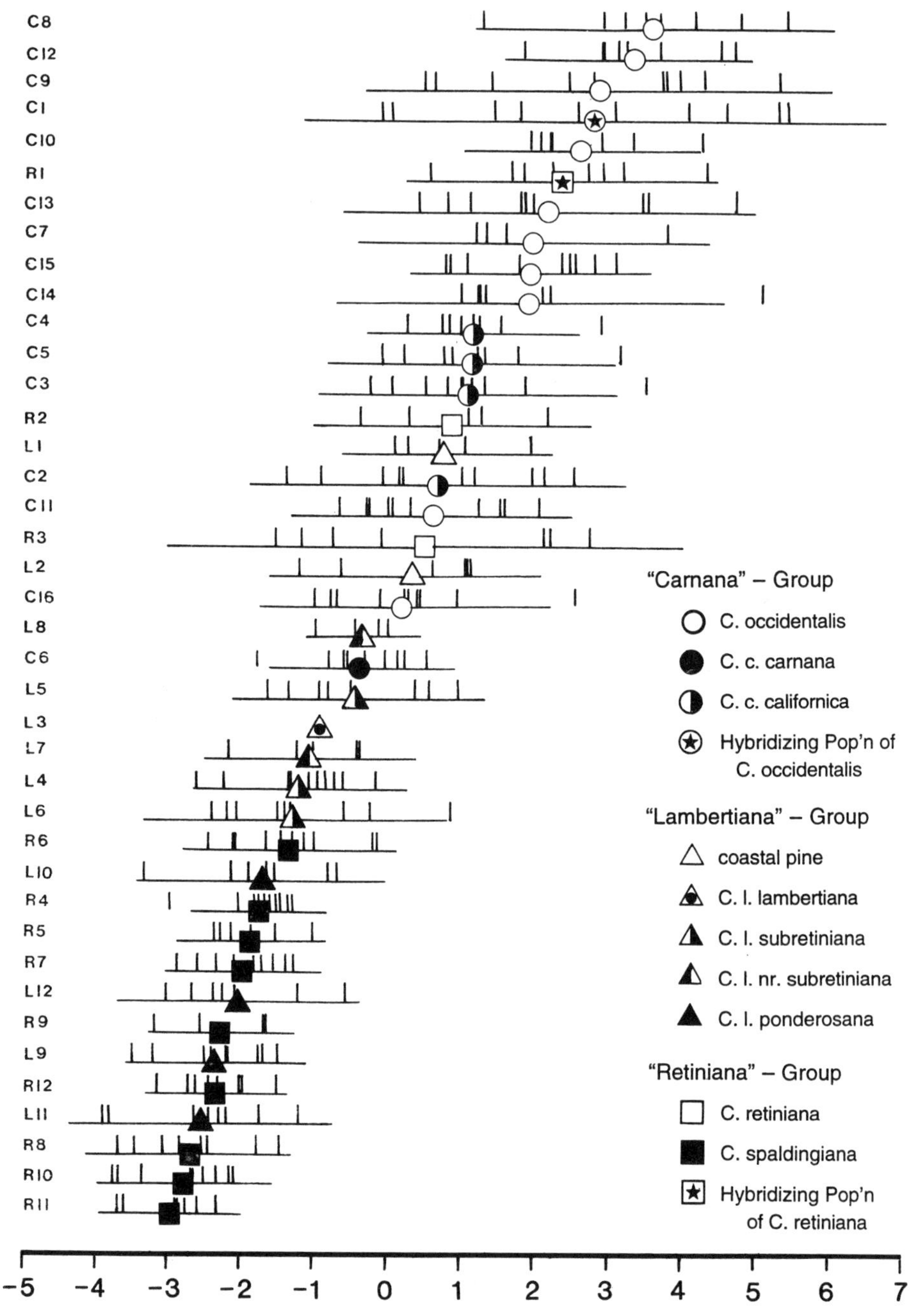

Fig. 4.12. First canonical variable of discriminant function analysis. The population symbol indicates the mean discriminant score (centroid) for each, as abbreviated on the right. Short vertical bars are individual scores and horizontal lines are the 95% confidence intervals about the mean.

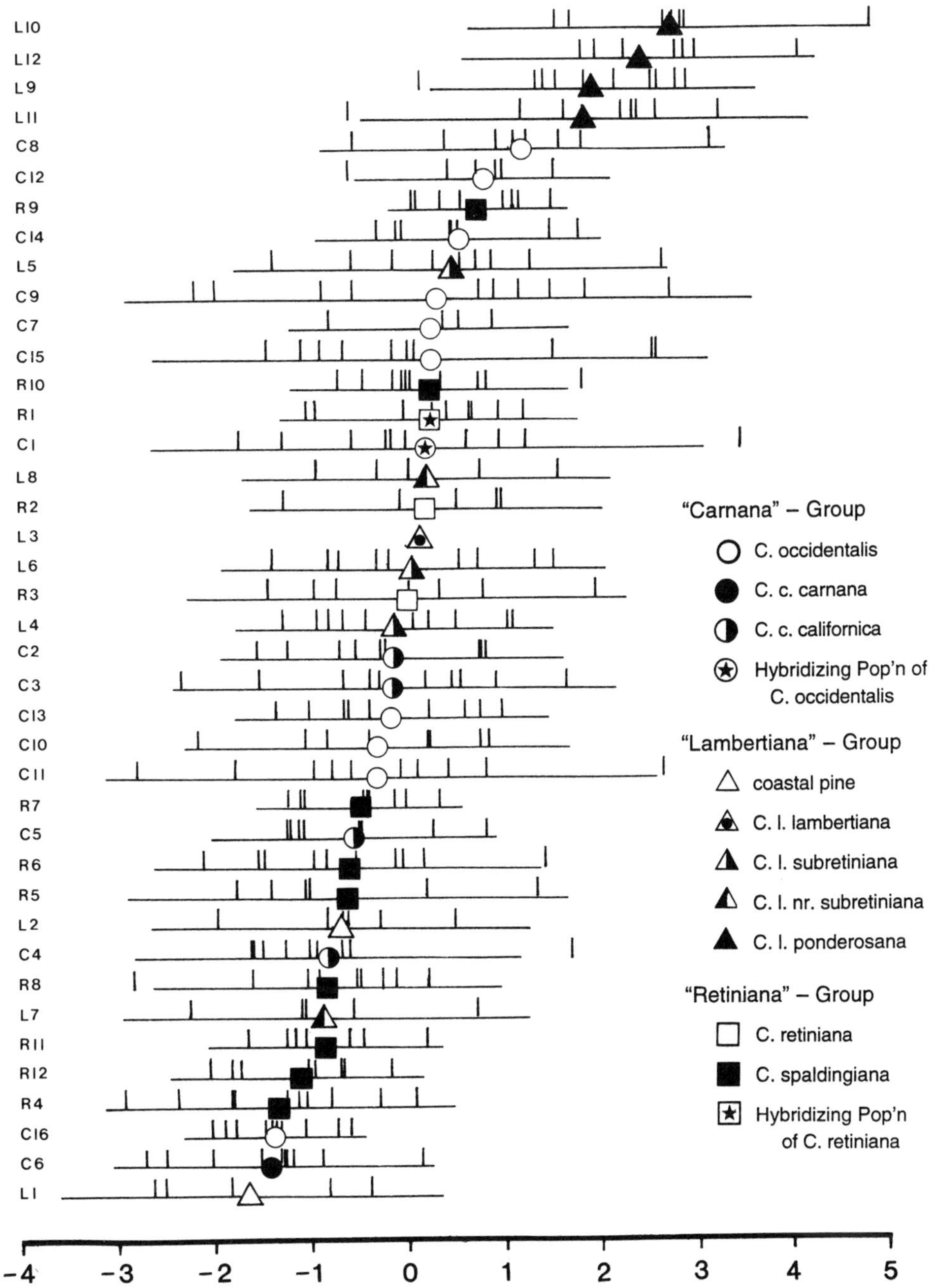

Fig. 4.13. Second canonical variable of discriminant function analysis. The population symbol indicates the mean discriminant score (centroid) for each, as abbreviated on the right. Short vertical bars are individual scores and horizontal lines are the 95% confidence intervals about the mean.

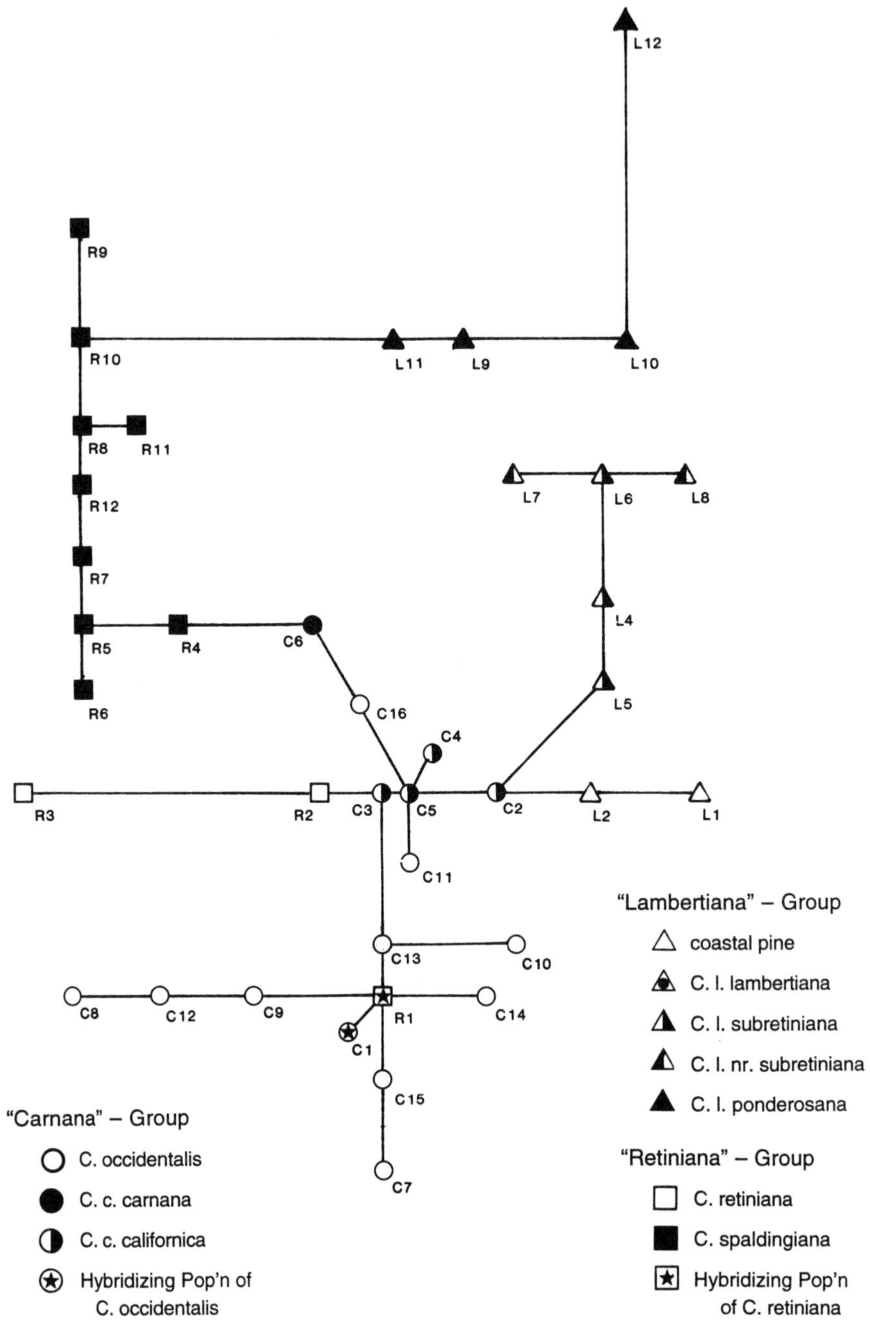

Fig. 4.14. Primnet based on discriminant function analysis. Each population is linked to its nearest neighbor by a line proportional in length to the Euclidean distance between their canonical variable centroids.

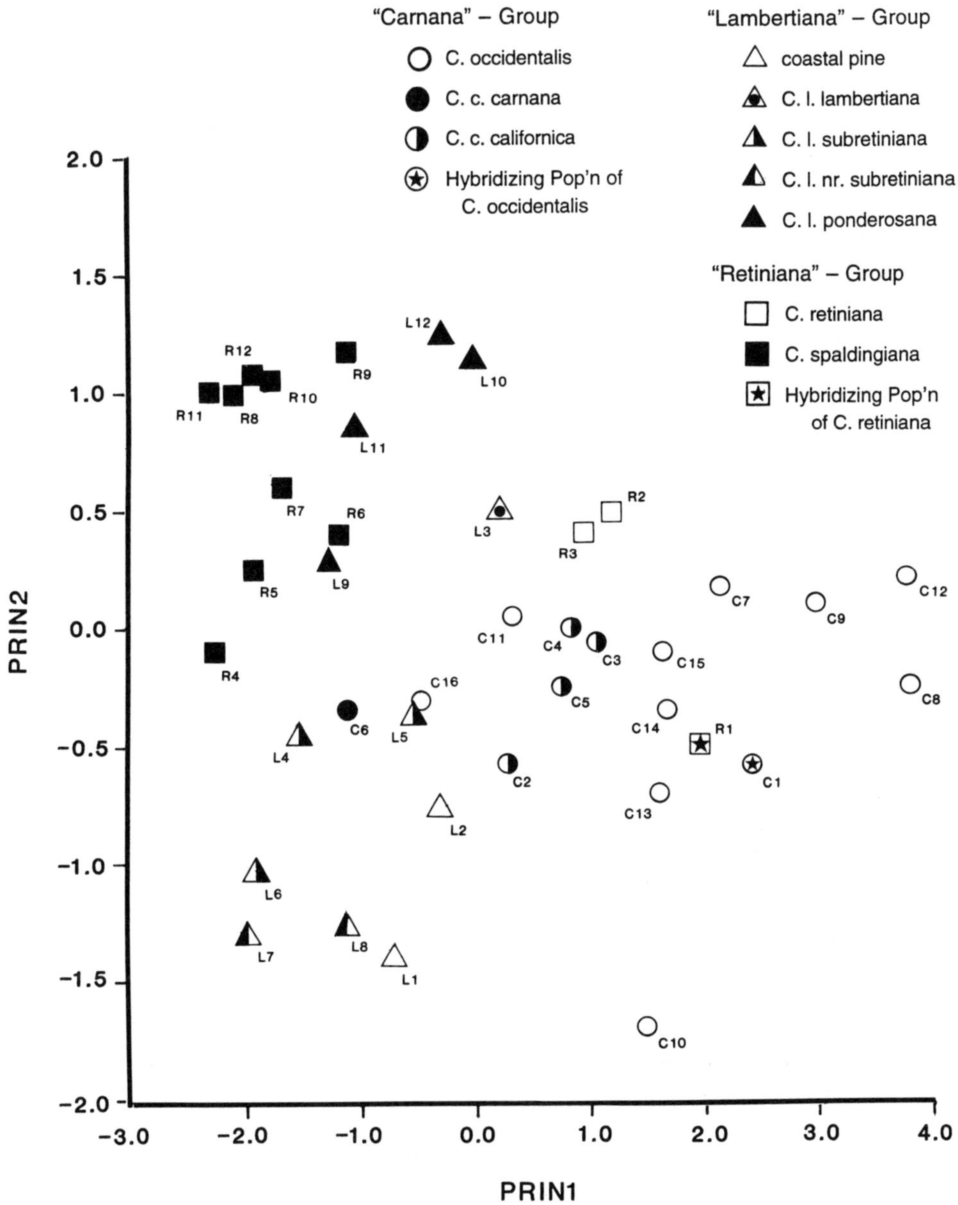

Fig. 4.15. Scattergram of population centroids of the first (PRIN1) and second (PRIN2) principal components.

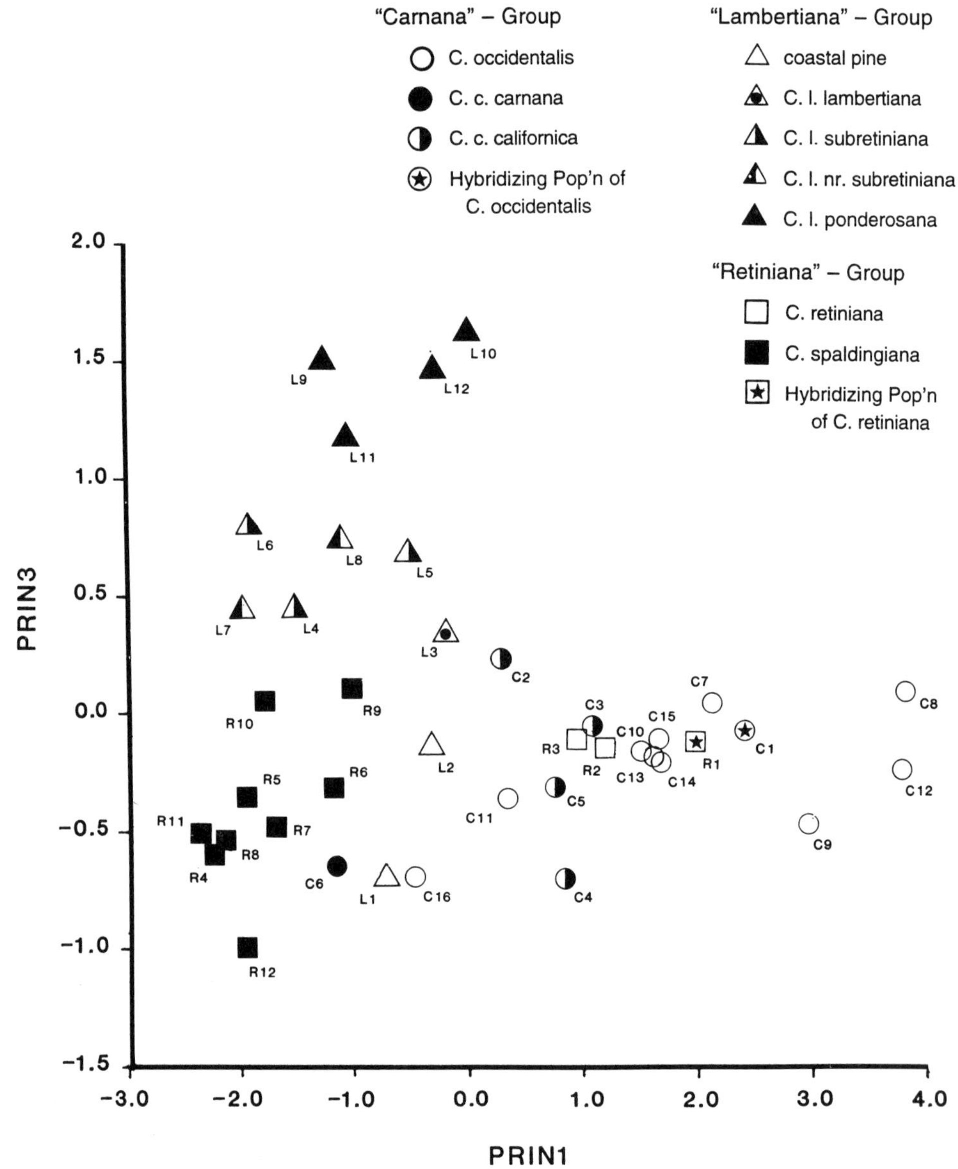

Fig. 4.16. Scattergram of population centroids of the first (PRIN1) and third (PRIN3) principal components.

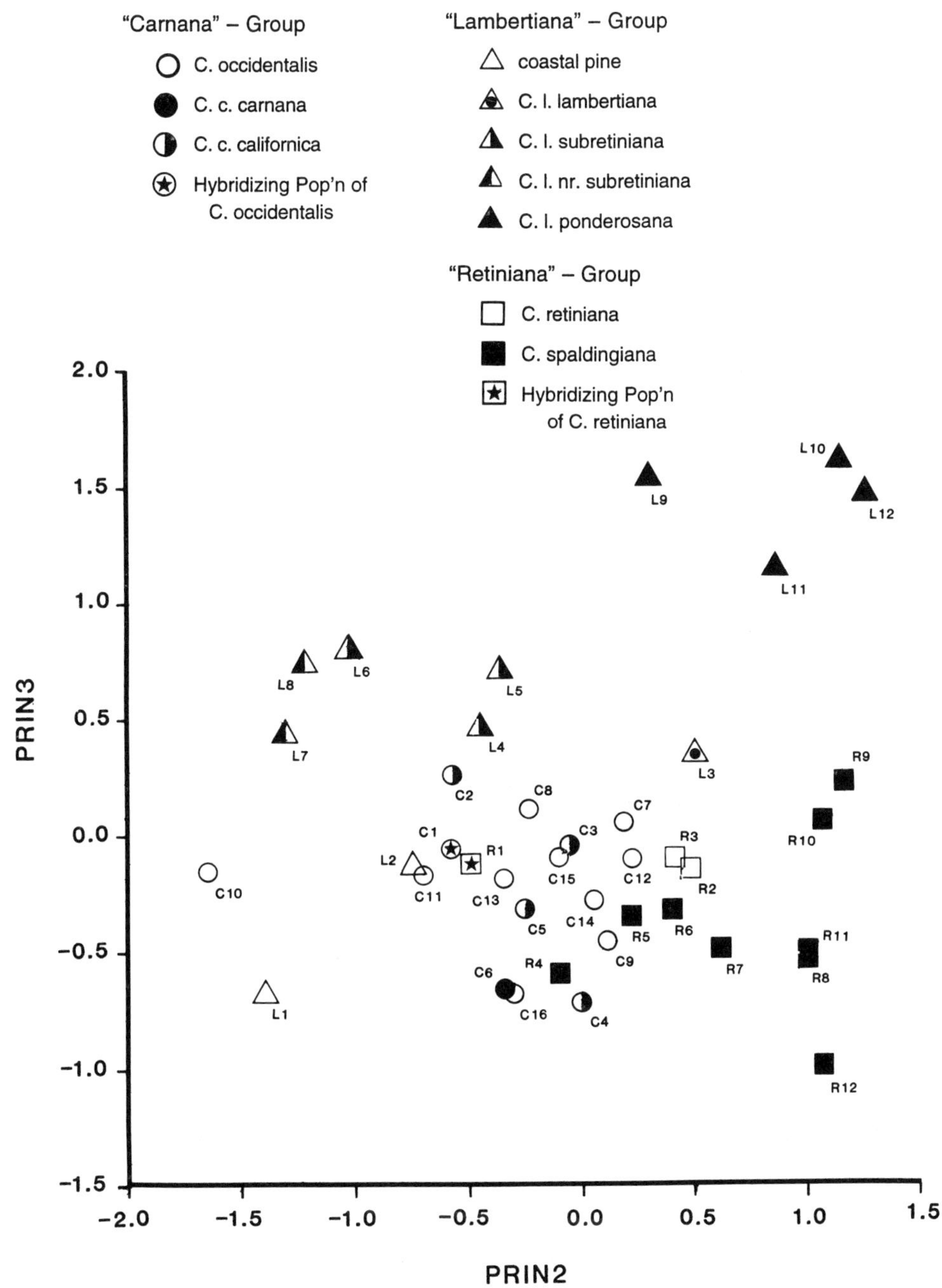

Fig. 4.17. Scattergram of population centroids of the second (PRIN2) and third (PRIN3) principal components.

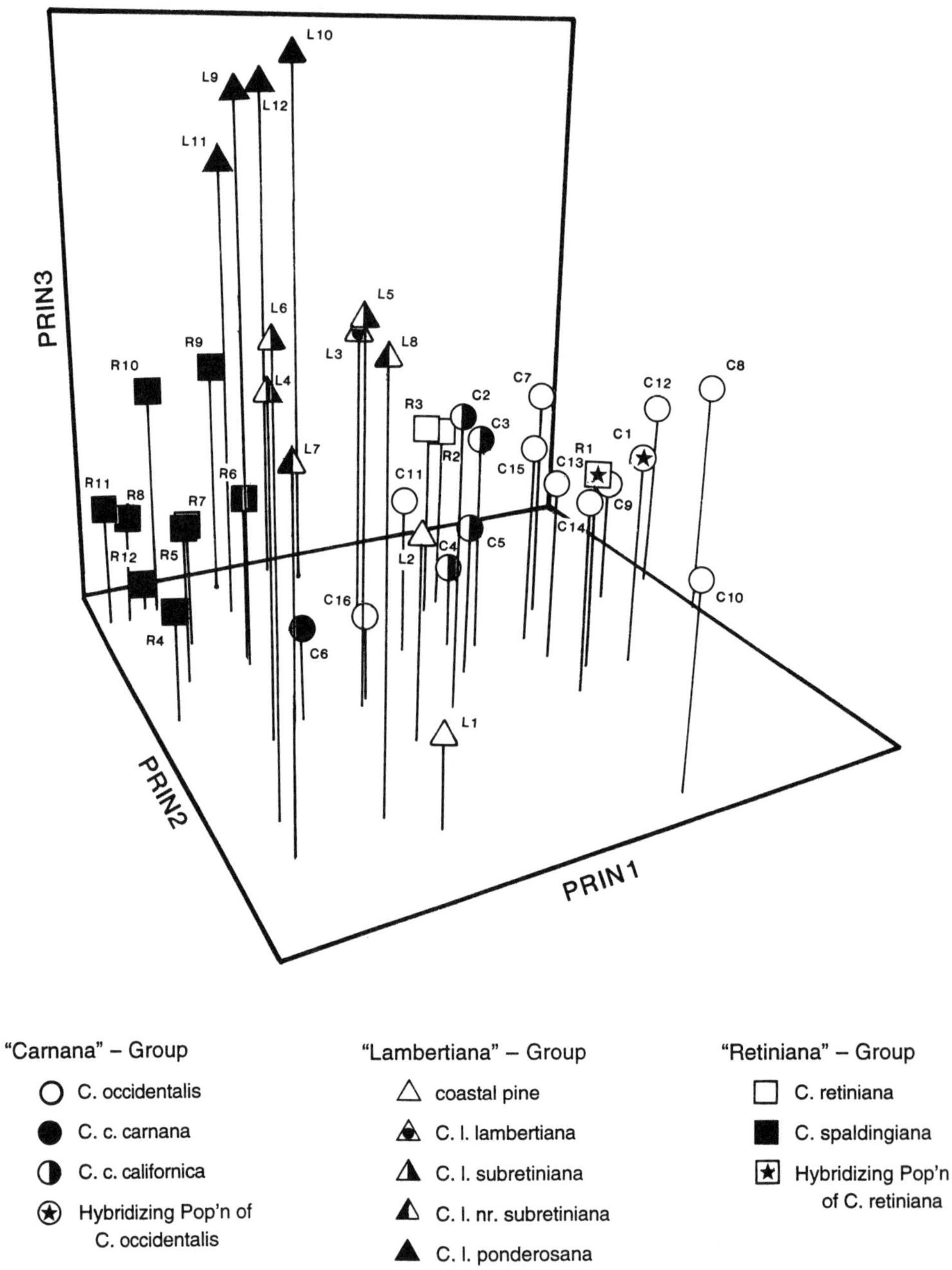

Fig. 4.18. Ball and stick diagram of the population centroids of the first three principal components, PRIN1, PRIN2 and PRIN3.

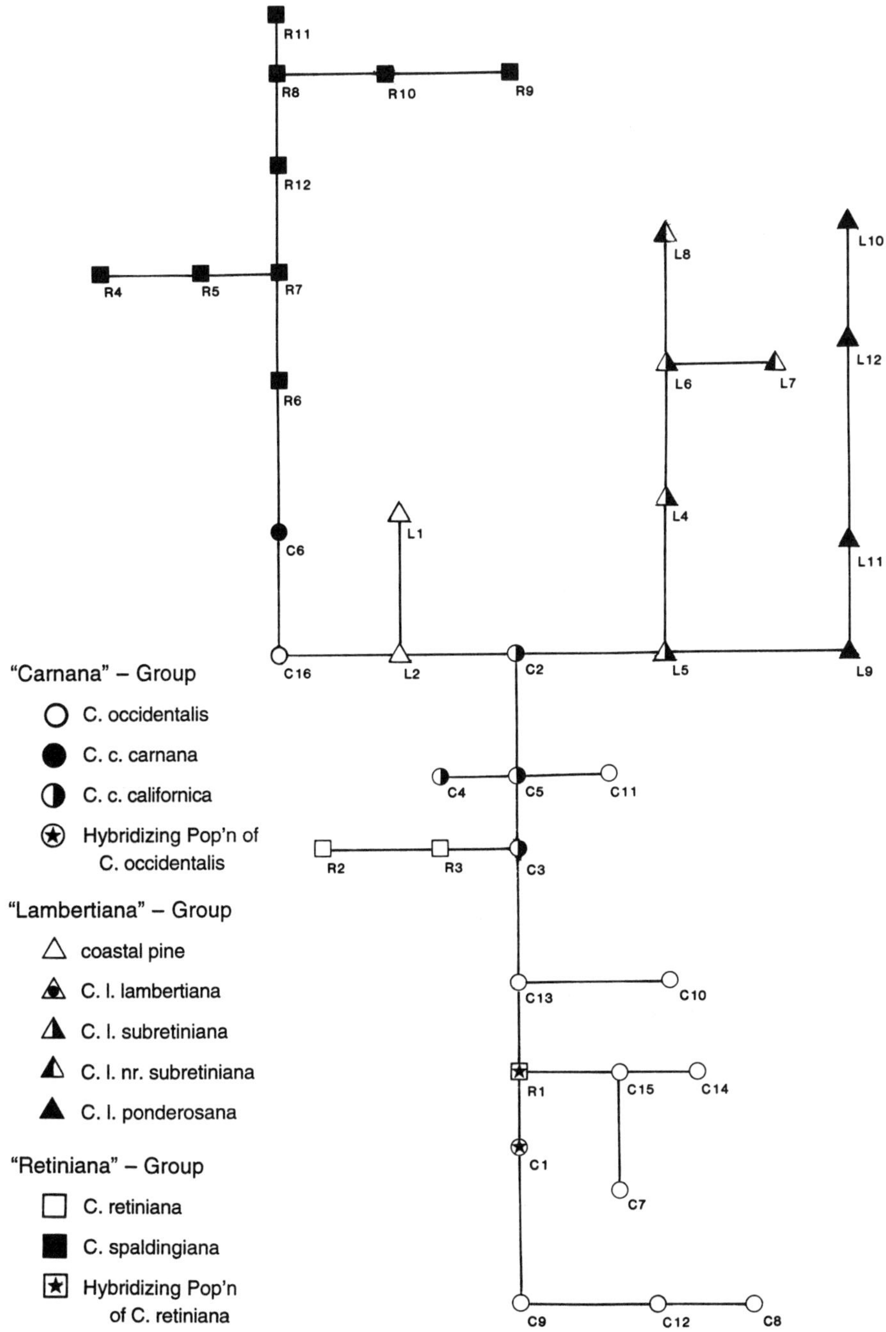

Fig. 4.19. Primnet based upon principal component analysis. Each population is linked to its nearest neighbor by lines proportional to the Euclidean distance between their principal components.

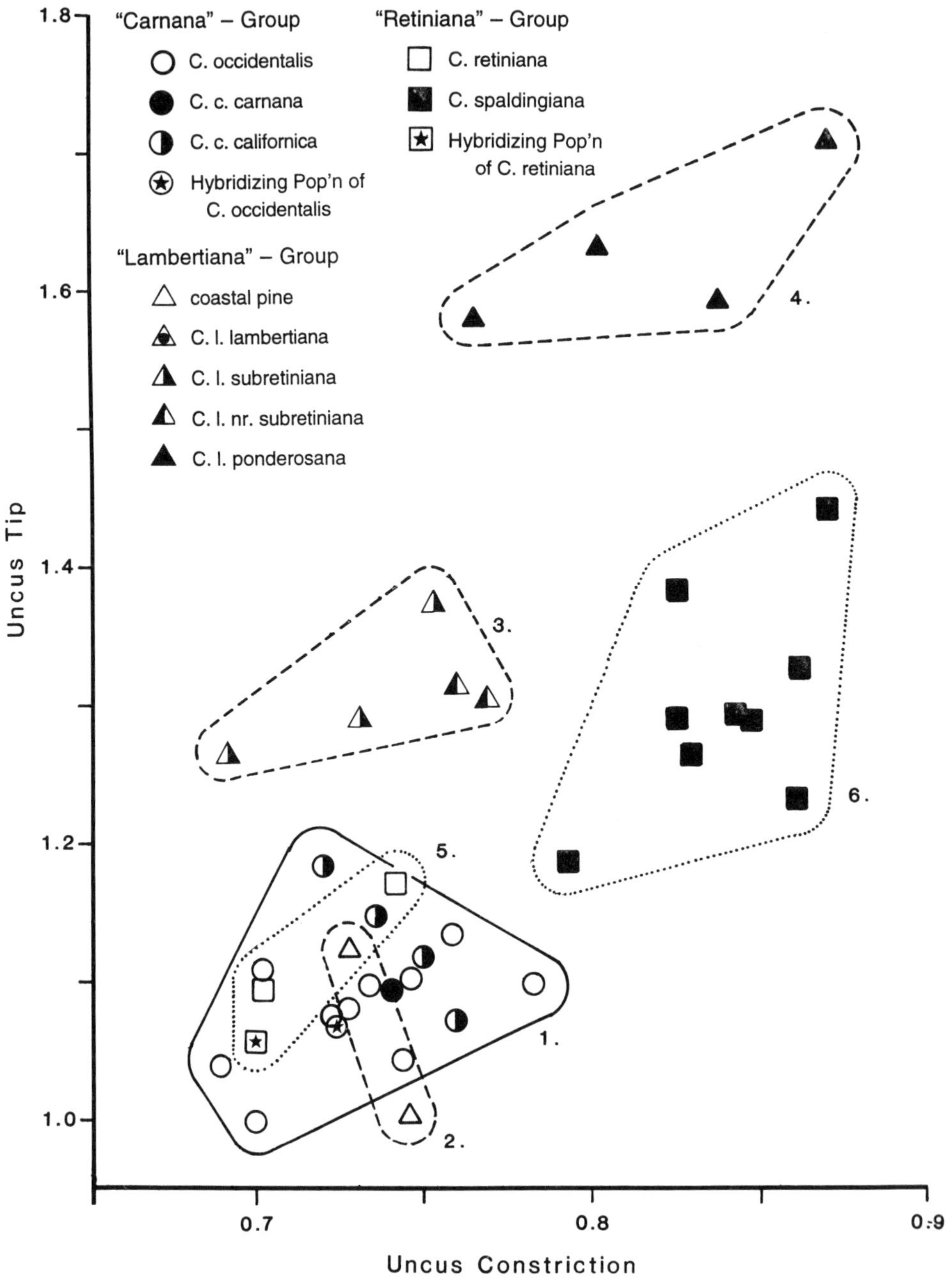

Fig. 4.20. Scattergram of uncus measurements by population adjusted for size. The average uncus tip width divided by basal width (y-axis) is plotted against minimum constriction width divided by basal width (x-axis). Cluster 1 = "occidentalis" group; 2 = coastal pine populations; 3 = *C. l. subretiniana* and *C. l.* nr. *subretiniana*; 4 = *C. l. ponderosana*; 5 = *C. retiniana*; and 6 = *C. spaldingiana*.

Appendix 4.1. Study Populations: Collection Data

The abbreviation for each population is followed by a summary of label data (excluding collectors' names), lot numbers of reared specimens in brackets, sources of specimens in parentheses (see key to institutions at end of appendix), sympatry of other *Choristoneura* subspecies, and whether chemical attraction was tested if not included with label locality data.

"Carnana" Group

C. occidentalis Freeman
(presumed hybridizing populations)

The other *C. occidentalis* populations are listed below (C7-C16).

C1: Grant Co., Oregon. Ten, drawn to OXY attractant, Starr Campground, 15 km SW Canyon City, 2-VIII-82 (UCB). Sympatry: *C. retiniana* (hybridizing population R1).

C. carnana californica Powell

C2: Siskiyou Co., California. Seven, reared on *Pseudotsuga menziesii*, Shadow Creek Campground, 8 km NE Cecilville, 23-VI-80, emerged 5 to 14-VII-80 [J. Powell no. 80F58]; 1, reared on same host, same locality, 20-VI-78 [W.J.A. Volney pt. 1, pupa 2]; 1, reared on same host, Etna Creek, 4 km SW Etna, 24-VI-80, emerged 11-VII-80 [J. Powell no. 80F73] (all UCB). Sympatry: None. Attraction confirmed 1980-81.

C3: Lake Co., California. Three, Anderson Springs, 9-VI-61 (UCD); 2 paratypes, same locality, 24-VI-58 (UCB); 3, 1 mi. NW Cobb, 26-VI-79 (UCB); 1, same locality, reared on *P. menziesii*, 5-VI-80, emerged 29-VI-80 [J. Powell no. 80F12] (UCB); 1, Hobergs, reared on same host, 27-V-80, emerged 11-VII-80 [J. Powell no. 80E49] (UCB). Sympatry: None. Attraction confirmed 1979-80.

C4: Plumas Co., California. One, Johnsville, 6-VI-60 (UCB); 1, same locality, 28-VII-68 (UCD); 1, 4 mi. W Quincy, 24-VI-49 (UCB); 3, Plumas-Eureka State Park, 27-VII-68 (UCB); 4, 1 mi. W Johnsville, 17 and 18-VII-73 (UCD). Sympatry: *C. spaldingiana* (R4)(at some localities). Attraction confirmed 1979-81.

C5: El Dorado Co., California. Five, Blodgett Forest, 13 mi. E Georgetown, 13 to 17-VII-67; 3, same locality, 27-VII-78; 2, same locality, 12-VII-80 (all UCB). Sympatry: *C. spaldingiana* (R6). Attraction confirmed 1978-82.

C. carnana carnana (Barnes and Busck)

C6: San Bernardino Co., California. Nine, Manker Flat campground, 4 km N Mt. Baldy P.O., 18-VII-79; 1, Glacier Picnic Area, 3 km NE Mt. Baldy P.O., same date (all UCB). Sympatry: *C. spaldingiana* (R8). Attraction confirmed 1979-80.

C. occidentalis Freeman

C7: Missoula Co., Montana. Four, reared on Douglas-fir cones, Superior, 27-VI-68 [Hopkins no. 52067 (3), and 52046 (1)]; 3, reared on larch, Bonner, 17-VI-69 [Hopkins no. 54299]; 4, reared on lodgepole, same locality, 15 to 31-VI-69 [Hopkins no. 54295, 54303, 54326, and 54335] (all NMNH). Sympatry: *C. lambertiana* nr. *subretiniana* (L8). Attraction not tested.

C8: Lehman Co., Idaho. Ten, Twin Creek Camp, 5 mi. N Gibbonsville, 5200 ft., 29 to 31-VII-56 (AMNH). Sympatry: *C. lambertiana* nr. *subretiniana* (not examined). Attraction not tested.

C9: Elko Co., Nevada. Four, Pine Creek Campground, 5 km S Jarbidge, 6600 ft., 5-VIII-81; 6, Bear Creek Summit, 6 km SW Jarbidge, 6-VIII-81 (all UCB). Sympatry: None. Attraction confirmed 1981.

C10: Utah Co, Utah. Five, Timpooneche Campground, Mt. Timpanogos, 29-VII-67; 5, same locality, 24-VII-81 (all UCB). Sympatry: *C. spaldingiana* (R9). Attraction confirmed 1981.

C11: White Pine Co., Nevada. Eight, Timber Creek Campground, 13 km E McGill, 7-VIII-80; 2, Timber Creek, 10 km E McGill, 18-VIII-80 (all UCB). Sympatry: *C. spaldingiana* (not examined). Attraction confirmed 1980.

C12: White Pine County, Nevada. Eight, Lehman Creek, 7 mi. W Baker, 7 to 9-VIII-78 (UCB). Sympatry: *C. spaldingiana* (R10). Attraction not tested on moths of this phenotype. In 1980 moths of phenotype of C11 were drawn to OXY attractant at this locality when moths resembling the 1978 phenotype were not seen.

C13: Kane Co., Utah. Ten, Duck Creek, 5 km E Navajo Lake, 8500 ft., 23-VII-81 (UCB). Sympatry: *C. lambertiana ponderosana* (L9) and *C. spaldingiana* (R11). Attraction confirmed 1981.

C14: Montrose Co., Colorado. Ten, Iron Springs Camp, 25 mi. SW Montrose, 4 to 7-VIII-60 (AMNH). Sympatry: Unknown, possibly, *C. lambertiana ponderosana*. Attraction not tested.

C15: Coconino Co., Arizona. Ten, Hochderffer Hill, 12.5 mi. NNW Flagstaff, 16 to 18-VII-61 (NMNH). Sympatry: *C. lambertiana ponderosana* (L11). Attraction confirmed in 1981.

C16: Coconino Co., Arizona. Ten, West Fork, 16 mi. SW Flagstaff, 4 to 19-VII-61 (NMNH). Sympatry: Unknown, possibly *C. lambertiana ponderosana*. Attraction not tested.

"*Lambertiana*" Group

Coastal Pine-Feeding Populations

L1: Pacific Co., Washington. Ten, reared on *Pinus contorta*, Long Beach, 24-VII to 2-VIII-66 [Hopkins no. 33540-A] (Corvallis [8], NMNH [2]). Sympatry: Unknown. Attraction not tested.

L2: Coos Co., Oregon. Ten, Tugman State Park near Lakeside, 9 and 10-VIII-76 (UCB). Sympatry: None. Attraction not tested.

C. lambertiana lambertiana (Busck)

L3: Jackson Co., Oregon. Two (including a "cotype"), reared on *Pinus lambertiana*, Ashland, 7-VI and 1-VII-14 (NMNH). Sympatry: Unknown at time of collection. In 1982, moths on Mt. Ashland were drawn to the three attractants. Those drawn to the SUBRET attractant did not resemble this phenotype. There was no *P. lambertiana* at the trapping site.

C. lambertiana subretiniana Obraztsov

L4: Modoc Co., California. Ten, Warner Mts., 3 mi. E Davis Creek, 24-VII to 7-VIII-22 (NMNH). Sympatry: *C. retiniana* (R2). Attraction not tested.

L5: Placer Co., California. Five, Ward Creek, 2 mi. S Tahoe City, 6050 ft., 20-VII to 6-VIII-66; 2, same locality, 23-VII and 12-VIII-74; 3, same locality, 27-VII to 14-VIII-75 (all LACM). Sympatry: *C. spaldingiana* (R5). Attraction confirmed in 1980.

L6: Tulare Co., California. Ten, Monache Meadows, 8000 ft., 8-VII to 15-VIII (no year on labels) (NMNH). Sympatry not known. Attraction not tested. The type series of *C. l. subretiniana* was drawn from this collection, but there were no designated paratypes among the specimens in the study sample.

C. lambertiana near *subretiniana*

L7: Missoula Co., Montana. Four, reared on lodgepole pine, Martin City, 11-VII-67 [Hopkins no. 51776] (NMNH); 3, same data except 19-VII-68 [Hopkins no. 52148] (UCB). Sympatry unknown, but *C. occidentalis* is probable. Attraction not tested.

L8: Missoula Co., Montana. Three, reared on lodgepole pine, Belmont Cr., Missoula, 7-VII-68 [Hopkins no. 52100] (UCB); 1, same data except 10-VI-69 [Hopkins no. 52108] (NMNH); 2, reared on lodgepole pine, Evaro, 25-VI and 10-VII-68 (UCB); 1, reared on lodgepole pine, Chamberlain Cr., Bonner, 17-VI-69 [no lot number] (NMNH). Sympatry: *C. occidentalis* (C7). Attraction not tested.

C. lambertiana ponderosana Obraztsov

L9: Kane Co., Utah. Three, drawn to blacklight, Duck Cr., 5 km. E Navajo Lake, 23-VII-81; 7, drawn to blacklight, Willis Cr., 10 km. E Navajo Lake, 23-VII-81 (all UCB). Sympatry: *C. occidentalis* (C13) and *C. spaldingiana* (R11). Attraction confirmed in 1981.

L10: Clark Co., Nevada. Ten, drawn to blacklight, McWilliams Campground, Charleston Mts., 20-VII-81 (UCB). Sympatry: *C. spaldingiana* (R12). Attraction confirmed in 1981.

L11: Coconino Co., Arizona. One, Fort Valley, 7.5 mi. NW Flagstaff, 6-VII-61 (NMNH); 1, Hart Prairie, 10 mi. NW Flagstaff, 14-VII-61 (NMNH); 4, drawn to blacklight, Hochderffer Hill, 20 km. NE Flagstaff, 21-VII-81 (UCB); 3, same locality, drawn to SUBRET bait, 22-VII-81 (UCB); 1, Mt. Elden, 7 km. NE Flagstaff, 22-VII-81 (UCB). Sympatry: *C. occidentalis* (C15). Attraction confirmed in 1981.

L12: Apache Co., Arizona. Nine, Alpine Divide Camp, 4 mi. N Alpine, 15 to 18-VII-65 (AMNH). Sympatry unknown. Attraction not tested.

"*Retiniana*" Group

C. retiniana (Walsingham)

(hybridizing population)

R1: Grant Co., Oregon. Ten, drawn to RET bait, same locality and same date as population C1. Sympatry: *C. occidentalis* (hybridizing population C1).

C. retiniana (Walsingham)

R2: Modoc Co., California. Nine, reared on *Abies concolor* as follows: Two, 17 mi. S Cedar Pass, 24-VI-74, emerged 26-VI and 1-VII-74 [J. Powell no. 74F72]; 2, 10 mi. N Cedar Pass, 7-VII-74, emerged 27-VII-74 [J. Powell no. 74G32]; 1, from 6 mi. NW Cedar Pass, 30-VI-74, emerged 28 and 30-VI-74 [J. Powell no. 74G16]; 2, 2 mi. NW Fandango Pass, 4-VII-74, emerged 26-VII and 8-VIII-74 [J. Powell no. 74G30]; 1, Cedar Mtn., 23-V-74, emerged 28-VI-74 [J. Powell no. 74F67]; 1, 10 mi. E Adin, 2-VII-74, emerged 14-VII-74 [J. Powell no. 74G28]. Lassen Co., CA: One, reared on *A. concolor*, 14 mi. NE Madeline, 1-VII-74, emerged 27-VII-74 [J. Powell no. 74G24] (all UCB). Sympatry: *C. lambertiana subretiniana* (L4). Attraction confirmed 1981.

R3: Siskiyou Co., California. One, 5 mi. E McCloud, 7-VII-57; 3, McBride Springs, Mt. Shasta, 21-VII-66; 2, same locality, reared on *A. concolor*, 10/14-VI-74, emerged 2 and 10-VII-74 [J. Powell no. 74F31]; 2, Castle Lake, 20-VII-68; 1, reared on *A. concolor*, south side Mt. Shasta, 6,250 ft., 24-VI-74, emerged 15-VII-74 [J. Powell no. 74F56.7]; 1, reared on *A. concolor*, 9 mi. N Bartle, 12/15-VI-74, emerged 2-VII-74 [J. Powell no. 74F44] (all UCB). Sympatry: *C. carnana californica* (at most localities). Attraction confirmed 1981.

C. spaldingiana Obraztsov

R4: Plumas Co., California. One, Plumas-Eureka State Park, 27-VII-68 (UCB); 9, 1 mi. E Johnsville, 16 to 18-VII-73 (UCD). Sympatry: *C. carnana californica* (C4). Attraction confirmed 1979-80.

R5: Placer Co., California. One, Ward Creek, 2 mi. S Tahoe City, 13-VIII-66; 1, same locality, 13-VIII-73; 7, same locality, 23-VII to 6-VIII-74; 1, same locality, 6-VIII-

75 (all LACM). Sympatry: *C. lambertiana subretiniana* (L5). Attraction confirmed 1979-81.

R6: El Dorado Co., California. Three, Blodgett Forest, 13 mi. E Georgetown, 12 and 14-VII-67; 1, same locality, 19-VI-79, reared on *A. concolor*, emerged 14-VII-79 [J. Powell no. 79F30]; 6, same locality, 17 to 24-VII-80 (all UCB). Sympatry: *C. carnana californica* (C5). Attraction confirmed 1978-82.

R7: Kern Co., California. Ten, Tehachapi Mt. Park, 17-VII-79 (UCB). Sympatry: None. Attraction confirmed 1979-80.

R8: San Bernardino Co., California. Six, Manker Flat Campground, 4 km N Mt. Baldy P.O., 18-VII-79; 1, Glacier Picnic Area, 3 km E Mt. Baldy P.O., 18-VII-79. Los Angeles Co., CA. Three, Table Mt., 1 km N Big Pines, 20-VII-79 (all UCB). Sympatry: *C. carnana* (C6), except at last locality. Attraction confirmed 1979-1980.

R9: Utah Co., Utah. Ten, drawn to RET bait, Mt. Timpanogos, 25-VII-81 (UCB). Sympatry: *C. occidentalis* (C10).

R10: White Pine Co., Nevada. Ten, drawn to RET bait, Lehman Creek, 11 km W Baker, 25-VII-81 (UCB). Sympatry: *C. occidentalis* (C12).

R11: Kane Co., Utah. Ten, drawn to RET bait, Willis Creek, 10 km E Navajo Lake, 23-VII-81 (UCB). Sympatry: *C. occidentalis* (C13) and *C. lambertiana ponderosana* (L9).

R12: Clark Co., Nevada. Ten, McWilliams Campground, Charleston Mts., 20-VII-81 (UCB). Sympatry: *C. lambertiana ponderosana* (L11). Attraction confirmed 1981.

KEY TO COLLECTION ABBREVIATIONS

AMNH = American Museum of Natural History, New York, N.Y.

Corvallis = U.S. Forest Service Pacific Northwest Forest and Range Experiment Station, Corvallis, Oregon.

LACM = Los Angeles County Museum of Natural History, Los Angeles, Calif.

NMNH = U.S. National Museum of Natural History, Washington, D.C.

UCB = Essig Museum of Entomology, University of California, Berkeley.

UCD = University of California, Davis.

Appendix 4.2. Summary Statistics Of Morphological Characters

FOREWING LENGTH (mm)

	C1	C2	C3	C4	C5	C6
Mean	14.58	13.02	13.45	13.56	13.27	12.24
S.D.	1.88	0.94	0.69	0.61	0.73	0.61
Max.	17.42	13.94	14.47	14.39	14.09	13.18
Min.	11.21	11.21	12.27	12.73	11.82	11.36
n	10	10	10	10	10	10

	C7	C8	C9	C10	C11	C12
Mean	13.88	15.55	14.89	14.02	13.37	15.42
S.D.	1.19	0.64	1.82	0.66	1.07	0.81
Max.	16.36	16.97	17.20	14.85	14.85	16.36
Min.	12.42	14.47	12.42	12.58	11.21	14.09
n	11	10	10	10	10	8

	C13	C14	C15	C16
Mean	13.95	14.92	14.52	12.41
S.D.	0.64	0.56	1.12	0.38
Max.	14.70	16.06	17.12	13.03
Min.	12.88	14.24	13.26	11.52
n	10	10	10	10

	L1	L2	L3	L4	L5	L6
Mean	12.45	12.86	10.61	12.12	12.95	12.58
S.D.	0.84	0.60	1.71	0.85	0.69	0.45
Max.	13.94	13.79	11.82	13.33	13.86	12.58
Min.	11.06	11.97	9.39	10.61	11.82	11.29
n	10	10	2	10	12	10

	L7	L8	L9	L10	L11	L12
Mean	11.54	12.14	12.73	13.69	12.75	13.48
S.D.	0.50	0.86	0.44	0.58	0.55	0.51
Max.	12.20	13.03	13.26	14.62	13.33	14.55
Min.	10.76	10.61	11.74	12.65	11.52	12.88
n	7	7	10	10	9	9

	R1	R2	R3	R4	R5	R6
Mean	14.39	13.15	13.45	11.77	12.24	12.32
S.D.	1.24	1.00	0.98	0.52	0.45	0.44
Max.	16.59	14.70	14.85	12.50	13.03	13.18
Min.	12.42	10.91	11.74	10.68	11.52	11.52
n	10	10	10	10	10	10

	R7	R8	R9	R10	R11	R12
Mean	12.50	11.83	13.06	12.32	11.72	12.08
S.D.	0.34	0.79	0.41	0.44	0.51	0.43
Max.	13.03	12.58	13.79	13.18	12.42	12.95
Min.	12.05	9.85	12.58	11.82	10.61	11.52
n	10	10	10	10	10	10

FOREWING WIDTH (mm)

	C1	C2	C3	C4	C5	C6
Mean	6.20	5.42	5.67	5.78	5.61	5.02
S.D.	0.93	0.48	0.37	0.38	0.35	0.25
Max.	7.65	5.98	6.36	6.36	6.21	5.45
Min.	4.70	4.62	5.15	5.15	5.00	4.77
n	10	10	10	10	10	10

	C7	C8	C9	C10	C11	C12
Mean	5.89	6.55	6.18	5.78	5.52	6.55
S.D.	0.64	0.33	0.78	0.27	0.53	0.46
Max.	6.82	7.05	7.42	6.21	6.44	7.27
Min.	4.92	5.91	5.15	5.30	4.55	5.91
n	11	10	10	10	10	8

	C13	C14	C15	C16
Mean	5.78	6.14	6.17	5.20
S.D.	0.26	0.56	0.54	0.30
Max.	6.14	7.58	7.58	5.45
Min.	5.23	5.61	5.61	4.47
n	10	10	10	10

	L1	L2	L3	L4	L5	L6
Mean	4.99	5.23	4.66	4.83	5.23	4.89
S.D.	0.36	0.40	0.37	0.29	0.31	0.24
Max.	5.61	5.76	4.92	5.15	5.61	9.70
Min.	4.32	4.55	4.39	4.32	4.55	4.55
n	10	10	2	10	12	10

	L7	L8	L9	L10	L11	L12
Mean	4.83	5.01	4.92	5.36	4.81	5.57
S.D.	0.34	0.53	0.29	0.24	0.28	0.24
Max.	5.23	5.76	5.23	5.76	5.15	5.98
Min.	4.24	4.02	4.24	5.00	4.24	5.30
n	7	7	10	10	9	9

	R1	R2	R3	R4	R5	R6
Mean	5.98	5.20	5.50	4.70	4.90	5.10
S.D.	0.66	0.37	0.37	0.18	0.27	0.14
Max.	7.05	5.91	6.14	5.00	5.30	5.30
Min.	5.00	4.70	5.00	4.32	4.55	4.85
n	10	10	10	10	10	10

	R7	R8	R9	R10	R11	R12
Mean	4.92	4.69	5.10	4.86	4.55	4.82
S.D.	0.13	0.26	0.24	0.22	0.16	0.12
Max.	5.15	4.92	5.53	5.15	4.85	5.00
Min.	4.70	4.09	4.70	4.55	4.24	4.62
n	10	10	10	10	10	8

NUMBER OF MIDLEG BASITARSAL SPINES

	C1	C2	C3	C4	C5	C6
Mean	18.70	16.00	16.50	15.60	15.60	12.25
S.D.	3.40	2.58	3.34	2.95	2.91	1.98
Max.	26	20	23	20	22	15
Min.	15	12	13	11	11	9
n	10	10	10	10	10	10

	C7		C8		C9		C10	C11	C12
Mean	18.14		20.20		18.30		20.00	13.50	20.12
S.D.	1.86		3.43		3.34		3.56	2.80	1.89
Max.	20		27		23		26	17	23
Min.	16	15	13	16	9	17			
n	7	10	10	10	10	8			

	C13	C14	C15	C16
Mean	17.50	17.10	15.40	13.60
S.D.	3.27	3.11	2.37	4.08
Max.	23	22	18	20
Min.	13	13	11	9
n	10	10	10	10

	L1	L2	L3	L4	L5	L6
Mean	14.60	15.80	14.50	12.40	14.04	13.30
S.D.	2.80	2.15	4.95	1.65	2.24	2.26
Max.	18	19	18	15	18	16
Min.	10	12	11	11	10	9
n	10	10	2	10	12	10

	L7	L8	L9	L10	L11	L12
Mean	14.29	14.00	13.10	12.80	12.20	13.00
S.D.	1.60	2.58	1.79	2.10	3.39	2.55
Max.	17	18	15	17	20	17
Min.	13	11	9	11	9	10
n	7	7	10	10	10	9

	R1	R2	R3	R4	R5	R6
Mean	17.70	13.30	16.20	10.50	10.30	11.90
S.D.	2.83	3.16	4.21	1.35	2.11	2.68
Max.	20	10	24	12	13	18
Min.	12	9	10	8	7	7
n	10	10	10	10	10	10

	R7	R8	R9	R10	R11	R12
Mean	9.95	9.90	11.70	10.30	9.44	9.30
S.D.	1.54	1.79	2.98	1.64	1.42	1.06
Max.	12	12	16	13	12	11
Min.	7	6	8	8	8	8
n	10	10	10	10	9	10

NUMBER OF HINDLEG BASITARSAL SPINES

	C1	C2	C3	C4	C5	C6
Mean	19.30	16.40	16.50	14.70	15.45	12.45
S.D.	4.08	2.55	3.41	1.49	2.48	1.72
Max.	28	19	24	17	18.5	15
Min.	13	11	13	12	11	10
n	10	10	10	10	10	10

	C7	C8	C9	C10	C11	C12
Mean	18.57	21.70	19.60	19.10	15.10	19.88
S.D.	2.88	3.77	4.03	3.03	3.25	1.36
Max.	23	29	25	23	21	22
Min.	15	16	13	15	10	19
n	7	10	10	10	10	8

	C13	C14	C15	C16
Mean	17.80	16.50	16.40	14.40
S.D.	3.58	2.12	2.17	3.80
Max.	22	20	19	23
Min.	13	14	13	10
n	10	10	10	10

	L1	L2	L3	L4	L5	L6
Mean	15.50	16.80	15.50	13.60	14.50	13.70
S.D.	2.68	2.10	3.54	2.01	1.47	1.85
Max.	20	21	18	17	16	16
Min.	12	13	13	10	11	11
n	10	10	1	10	11	10

	L7	L8	L9	L10	L11	L12
Mean	14.00	14.57	13.30	14.30	12.90	12.75
S.D.	2.16	1.99	2.36	2.01	1.97	2.38
Max.	17	17	18	17	16	17
Min.	11	12	10	11	10	10
n	7	7	10	10	10	8

	R1	R2	R3	R4	R5	R6
Mean	17.90	13.44	14.70	9.80	9.40	11.40
S.D.	3.35	4.82	3.06	1.03	2.00	1.43
Max.	26	23	20	11	13	15
Min.	14	9	10	8	6	10
n	10	10	10	10	10	10

	R7	R8	R9	R10	R11	R12
Mean	9.30	8.50	9.60	9.00	7.50	7.70
S.D.	1.65	1.96	1.35	1.63	1.27	0.67
Max.	13	13	12	11	9	9
Min.	7	6	8	7	6	7
n	10	10	10	10	10	10

MAXIMUM WIDTH OF UNCUS BASE[1]

	C1	C2	C3	C4	C5	C6
Mean	6.94	6.31	6.66	6.64	6.54	6.16
S.D.	0.87	0.72	0.55	0.46	0.77	0.42
Max.	8.20	7.40	7.40	7.30	8.10	6.70
Min.	5.50	5.30	6.00	5.80	5.60	5.40
n	10	10	10	9	8	10

	C7	C8	C9	C10	C11	C12
Mean	6.99	7.19	7.33	6.55	6.65	7.56
S.D.	0.73	0.64	0.72	0.40	0.52	0.68
Max.	7.80	7.80	8.20	7.00	7.80	8.60
Min.	6.00	6.00	6.20	6.00	6.30	6.40
n	7	8	10	8	8	8

	C13	C14	C15	C16
Mean	6.43	6.94	6.85	6.19
S.D.	0.71	0.54	1.07	0.60
Max.	7.20	7.90	8.80	7.00
Min.	5.10	6.00	4.80	5.00
n	10	9	10	10

	L1	L2	L3	L4	L5	L6
Mean	6.08	6.23	5.60	5.68	6.04	5.13
S.D.	0.65	0.48	N/A	0.63	0.63	0.45
Max.	7.10	6.60	5.60	6.50	7.40	5.90
Min.	5.40	5.40	5.60	4.90	5.40	4.40
n	5	6	1	10	9	9

	L7	L8	L9	L10	L11	L12
Mean	5.18	5.24	5.20	5.80	5.33	5.26
S.D.	0.26	0.53	0.39	0.35	0.31	0.46
Max.	5.50	5.50	5.80	6.20	5.80	6.20
Min.	4.90	4.30	4.60	5.30	4.90	4.70
n	5	5	10	7	9	8

	R1	R2	R3	R4	R5	R6
Mean	6.92	7.17	6.76	5.61	5.40	5.68
S.D.	0.62	0.51	0.61	0.30	0.24	0.42
Max.	7.70	8.00	7.50	5.90	5.80	6.40
Min.	6.00	6.50	6.00	4.90	5.10	4.90
n	9	6	7	10	6	10

	R7	R8	R9	R10	R11	R12
Mean	5.69	5.63	5.84	5.44	5.71	5.79
S.D.	0.23	0.44	0.52	0.38	0.44	0.23
Max.	6.00	6.40	6.80	6.00	6.50	6.10
Min.	5.40	4.90	5.20	5.00	5.30	5.40
n	9	9	8	10	8	8

1. Widths of projected image in millimeters. To convert to actual widths, multiply by 0.044.

MINIMUM WIDTH OF UNCUS CONSTRICTION[1]

	C1	C2	C3	C4	C5	C6
Mean	4.97	4.50	4.90	5.06	4.89	4.54
S.D.	0.42	0.37	0.63	0.57	0.69	0.37
Max.	5.90	5.00	6.00	6.30	5.90	5.10
Min.	4.20	3.90	3.90	4.10	3.80	3.90
n	10	10	10	9	8	10

	C7	C8	C9	C10	C11	C12
Mean	5.11	5.35	5.45	4.59	4.83	5.44
S.D.	0.60	0.32	0.59	0.40	0.32	0.39
Max.	5.70	5.70	6.40	5.00	5.10	6.10
Min.	4.00	5.00	4.60	4.10	4.10	5.10
n	7	8	10	8	8	8

	C13	C14	C15	C16
Mean	4.85	4.77	4.79	4.83
S.D.	0.55	0.33	0.84	0.46
Max.	5.90	5.10	6.80	5.70
Min.	4.00	4.20	3.90	4.20
n	10	9	10	10

	L1	L2	L3	L4	L5	L6
Mean	4.48	4.53	5.50	4.13	4.16	3.83
S.D.	0.40	0.66	N/A	0.37	0.33	0.39
Max.	4.90	5.30	5.50	4.90	4.80	4.60
Min.	3.90	3.50	5.50	3.60	3.70	3.20
n	5	6	1	10	9	9

	L7	L8	L9	L10	L11	L12
Mean	3.98	3.98	4.16	4.43	4.47	4.58
S.D.	0.32	0.50	0.39	0.51	0.44	0.42
Max.	4.40	4.60	4.80	5.10	5.30	5.10
Min.	3.50	3.30	3.70	3.70	3.90	3.90
n	5	5	10	7	9	8

	R1	R2	R3	R4	R5	R6
Mean	4.86	4.98	5.01	4.44	4.55	4.81
S.D.	0.71	0.60	0.51	0.23	0.15	0.40
Max.	5.90	5.90	5.60	4.70	4.70	5.50
Min.	3.90	4.10	4.30	4.10	4.30	4.00
n	9	6	7	10	6	10

	R7	R8	R9	R10	R11	R12
Mean	4.72	4.86	4.79	4.73	4.70	4.99
S.D.	0.23	0.39	0.25	0.29	0.26	0.41
Max.	5.10	5.50	5.00	5.40	5.10	5.70
Min.	4.40	4.10	4.20	4.30	4.40	4.50
n	9	9	8	10	8	8

1. Widths of projected image in millimeters. To convert to actual widths, multiply by 0.044.

MAXIMUM WIDTH OF UNCUS TIP[1]

	C1	C2	C3	C4	C5	C6
Mean	7.37	7.39	7.62	7.10	7.24	6.66
S.D.	1.03	0.60	0.70	0.71	0.50	0.63
Max.	10.00	8.30	8.70	8.10	7.80	8.30
Min.	6.50	6.20	6.70	6.00	6.40	6.10
n	10	10	10	9	8	10

	C7	C8	C9	C10	C11	C12
Mean	7.64	7.88	7.67	6.55	7.15	8.08
S.D.	1.04	0.73	1.13	0.72	0.44	0.63
Max.	8.80	8.70	9.30	7.70	8.10	8.90
Min.	5.90	6.40	5.60	5.50	6.70	7.00
n	7	8	10	8	8	8

	C13	C14	C15	C16
Mean	7.29	7.18	7.42	6.80
S.D.	1.07	0.58	1.36	0.89
Max.	9.10	8.00	10.70	8.30
Min.	5.30	6.10	6.10	5.60
n	10	9	10	10

	L1	L2	L3	L4	L5	L6
Mean	6.10	6.98	8.40	7.30	7.61	7.03
S.D.	0.48	0.50	N/A	0.92	0.82	1.02
Max.	6.60	7.50	8.40	8.40	8.90	8.60
Min.	5.60	6.40	8.40	5.90	6.00	5.90
n	5	6	1	10	9	9

	L7	L8	L9	L10	L11	L12
Mean	6.74	6.94	8.46	9.16	8.50	8.96
S.D.	0.61	1.12	0.44	0.95	1.04	0.63
Max.	7.70	8.50	9.00	11.10	10.60	10.00
Min.	6.00	5.50	7.70	8.20	7.30	8.00
n	5	5	10	7	9	8

	R1	R2	R3	R4	R5	R6
Mean	7.31	7.78	7.89	6.65	6.97	7.30
S.D.	0.80	0.79	0.83	0.59	0.62	0.73
Max.	9.00	8.70	9.50	7.70	7.80	8.50
Min.	5.90	6.90	7.00	5.90	6.30	6.40
n	9	6	7	10	6	10

	R7	R8	R9	R10	R11	R12
Mean	7.19	7.44	8.05	7.85	7.35	7.14
S.D.	0.44	0.29	0.48	0.86	0.66	0.65
Max.	7.80	8.00	8.90	9.90	8.50	8.20
Min.	6.50	7.00	7.50	7.00	6.40	6.00
n	9	9	8	10	8	8

1. Widths of projected image in millimeters. To convert to actual widths, multiply by 0.044.

NUMBER OF CORNUTI

	C1	C2	C3	C4	C5	C6
Mean	4.00	3.67	3.38	3.67	3.50	3.80
S..	1.76	0.71	1.41	1.00	1.20	0.84
Max.	7	4	6	5	6	5
Min.	2	3	2	2	3	3
n	10	9	8	9	8	5

	C7	C8	C9	C10	C11	C12
Mean	3.71	4.00	4.25	3.13	3.22	3.88
S.D.	1.11	0.82	0.89	0.64	0.97	0.83
Max.	6	5	6	4	5	5
Min.	3	3	3	2	2	3
n	7	7	8	8	9	8

	C13	C14	C15	C16
Mean	3.33	3.11	3.00	2.89
S.D.	0.87	1.05	0.94	0.60
Max.	5	4	5	4
Min.	2	2	2	2
n	9	9	10	9

	L1	L2	L3	L4	L5	L6
Mean	3.67	3.14	4.00	3.60	3.40	2.50
S.D.	0.52	0.90	N/A	0.84	0.89	0.84
Max.	4	4	4	5	4	4
Min.	3	2	4	2	2	2
n	6	7	1	10	5	6

	L7	L8	L9	L10	L11	L12
Mean	3.00	3.17	2.75	3.29	3.50	2.50
S.D.	0.89	1.17	0.71	0.76	0.53	0.53
Max.	4	5	4	4	4	3
Min.	2	2	2	2	3	2
n	6	6	8	7	8	8

	R1	R2	R3	R4	R5	R6
Mean	3.00	4.63	3.25	2.75	3.57	3.63
S.D.	0.53	0.92	0.71	0.46	0.79	0.92
Max.	4	6	5	3	5	5
Min.	3	4	3	2	3	3
n	8	8	8	8	7	8

	R7	R8	R9	R10	R11	R12
Mean	4.00	4.14	3.63	4.33	3.33	3.43
S.D.	1.41	1.57	0.74	0.82	1.12	0.53
Max.	5	7	4	6	5	4
Min.	2	2	2	4	2	3
n	5	7	8	6	9	7

PROPORTION OF UNMARKED, TAN FOREWING PHENOTYPE
(ALL MALES FROM COLLECTION SITE)
(Percent of Population)

	R1	R2	R3	R4	R5	R6
Tan	0.0	9.6	20.0	70.4	68.1	76.7
Other	100.0	90.4	80.0	29.6	31.9	23.3

	R7	R8	R9	R10	R11	R12
Tan	78.7	81.8	43.3	100.0	78.9	42.1
Other	21.3	18.2	56.7	0.0	21.1	57.9

Appendix 4.3. Diagnostic Characters

The first line of each block of data is a size category, the second is the mean ± the standard deviation, the third is the range, and the fourth is the number of individuals in the group of populations comprising the subspecies.

"CARNANA" GROUP

Subspecies	Forewing Length, mm	Forewing Breadth (Length ÷ Width)	Hindleg Basitarsal Spines
carnana (sens. str.) (C6)	short 12.2 ± 0.6 (11.4-13.2) n=10	moderately broad 2.44 ± 0.10 (2.32-2.59) n=10	moderately sparse 12.4 ± 1.7 (10-15) n=10
californica (sens. lat.) (C2-C5)	intermediate 13.3 ± 0.8 (11.2-14.5) n=40	broad 2.39 ± 0.11 (2.18-2.65) n=40	intermediate 5.8 ± 2.6 (11-24) n=40
californica (Klamath and Coast Ranges) (C2-C3)	intermediate 13.2 ± 0.8 (11.2-14.5) n=20	broad 2.39 ± 0.12 (2.18-2.65) n=20	moderately spinose 16.4 ± 2.9 (11-24) n=20
californica(?) (Sierra Nevada) (C4-C5)	intermediate 13.4 ± 0.7 (11.9-14.4) n=20	broad 2.36 ± 0.11 (2.21-2.59) n=20	intermediate 15.1 ± 2.0 (11-18) n=20
occidentalis (sens. lat.) (C7-C16)	long 14.3 ± 1.3 (11.2-17.2) n=99	broad 2.40 ± 0.10 (2.12-2.60) n=99	spinose 17.8 ± 3.7 (10-29) n=95
occidentalis (typical) (C7-C10, C12-C15)	long 14.7 ± 1.16 (12.5-17.2) n=79	broad 2.39 ± 0.10 (2.12-2.60) n=79	spinose 18.7 ± 3.4 (13-29) n=75

CARNANA GROUP cont.

Subspecies	Forewing Length, mm	Forewing Breadth (Length ÷ Width)	Hindleg Basitarsal Spines
occidentalis(?) (eastern Nevada) (C11)	intermediate 13.4 ± 1.1 (11.2-14.8) n=10	moderately broad 2.42 ± 0.09 (2.31-2.58) n=10	intermediate 15.1 ± 3.2 (10-21) n=10
occidentalis(?) (Arizona "carnana"-like) (C16)	short 12.4 ± 0.4 (11.5-13.0) n=10	broad 2.39 ± 0.09 (2.29-2.58) n=10	intermediate 14.4 ± 3.8 (10-23) n=10
occidentalis(?) (Oregon hybrid) (C1)	long 14.6 ± 1.9 (11.2-17.4) n=10	broad 2.36 ± 0.07 (2.28-2.51) n=10	spinose 19.3 ± 4.1 (13-28) n=10

Subspecies	Uncus Constriction (Min. Width ÷ Max. Basal Width)	Uncus Tip (Max. Tip Width ÷ Max. Basal Width)
carnana (sens. str.) (C6)	normal 0.74 ± 0.07 (0.60-0.83) n=10	normal 1.08 ± 0.09 (1.00-1.26) n=10
californica (sens. lat.) (C2-C5)	normal 0.74 ± 0.07 (0.57-0.87) n=37	normal 1.13 ± 0.14 (0.93-1.42) n=37
californica (Klamath and Coast Ranges) (C2-C3)	moderately wide 0.73 ± 0.08 (0.61-0.87) n=20	normal 1.17 ± 0.14 (0.96-1.42) n=20

CARNANA GROUP cont.

Subspecies	Uncus Constriction (Min. Width ÷ Max. Basal Width)	Uncus Tip (Max. Tip Width ÷ Max. Basal Width)
californica(?) (Sierra Nevada) (C4-C5)	normal 0.76 ± 0.08 (0.57-0.86) n=17	normal 1.09 ± 0.13 (0.93-1.40) n=17
occidentalis (sens. lat.) (C7-C16)	normal 0.73 ± 0.07 (0.58-0.92) n=88	normal 1.08 ± 0.12 (0.77-1.44) n=88
occidentalis (typical) (C7-C10, C12-C15)	normal 0.72 ± 0.06 (0.59-0.88) n=68	normal 1.07 ± 0.12 (0.77-1.44) n=68
occidentalis(?) (eastern Nevada) (C11)	moderately wide 0.76 ± 0.09 (0.58-0.86) n=10	normal 1.14 ± 0.12 (0.98-1.32) n=10
occidentalis(?) (Arizona "carnana-like) (C16)	moderately wide 0.78 ± 0.06 (0.68-0.92) n=10	normal 1.10 ± 0.08 (0.90-1.21) n=10
occidentalis(?) (Oregon hybrid) (C1)	normal 0.72 ± 0.09 (0.62-0.95) n=10	normal 1.07 ± 0.12 (0.90-1.22) n=10

"LAMBERTIANA" GROUP

Subspecies	Forewing Length, mm	Forewing Breadth (Length÷ Width)	Hindleg Basitarsal Spines
lambertiana (L3)	short 10.6 ± 1.7 (9.4-11.8) n=2	broad 2.27 ± 0.18 (2.14-2.40) n=2	intermediate 15.5 ± 3.5 (13-18) n=2
subretiniana (sens. lat.) (L4-L8)	short 12.2 ± 0.8 (10.6-13.9) n=46	moderately narrow 2.47 ± 0.11 (2.26-2.77) n=46	intermediate 14.1 ± 1.9 (10-17) n=45
subretiniana (sens. str.) (L4-L6)	short 12.4 ± 0.8 (10.6-13.9) n=32	moderately narrow 2.49 ± 0.09 (2.34-2.77) n=32	intermediate 14.0 ± 1.8 (10-17) n=31
near subretiniana (L7-L8)	short 11.9 ± 0.7 (10.6-13.1) n=14	moderately broad 2.42 ± 0.13 (2.26-2.64) n=14	intermediate 14.3 ± 2.0 (12-17) n=14
ponderosana (L9-L12)	intermediate 13.2 ± 0.7 (11.5-14.6) n=38	narrow 2.56 ± 0.12 (2.34-2.77) n=38	moderately sparse 13.3 ± 2.2 (10-18) n=38
coastal Oregon/Washington (L1-L2)	moderately short 12.7 ± 0.5 (11.5-13.9) n=20	moderately narrow 2.48 ± 0.09 (2.30-2.65) n=20	moderately spinose 16.2 ± 1.2 (12-21) n=20

"LAMBERTIANA" GROUP cont.

Subspecies	Uncus Constriction (Min. Width ÷ Max. Basal Width)	Uncus Tip (Max. Tip Width ÷ Max. Basal Width)
lambertiana (L3)	wide 0.98 (0.98) n=1	broad 1.5 (1.5) n=1
subretiniana (sens. lat.) (L4-L8)	normal 0.74 ± 0.08 (0.62-0.96) n=38	moderately broad 1.31 ± 0.15 (1.00-1.62) n=38
subretiniana (sens. str.) (L4-L6)	normal 0.72 ± 0.08 (0.62-0.96) n=28	moderately broad 1.31 ± 0.16 (1.00-1.62) n=28
near subretiniana (L7-L8)	moderately wide 0.76 ± 0.05 (0.67-0.84) n=10	moderately broad 1.32 ± 0.15 (1.09-1.57) n=10
ponderosana (L9-L12)	wide 0.82 ± 0.08 (0.60-0.96) n=34	broad 1.63 ± 0.16 (1.26-1.92) n=34
coastal Oregon/Washington (L1-L2)	normal 0.74 ± 0.10 (0.59-0.91) n=11	normal 1.08 ± 0.19 (0.79-1.27) n=11

"RETINIANA"GROUP

Subspecies	Forewing Length, mm	Forewing Breadth (Length ÷ Width)	Hindleg Basitarsal Spines
retiniana (R2-R3)	intermediate 13.3 ± 1.0 (10.9-14.8) n=20	moderately narrow 2.49 ± 0.12 (2.28-2.71) n=20	intermediate 14.1 ± 3.9 (9-23) n=19
retiniana(?) (Oregon hybrid) (R1)	long 14.4 ± 1.2 (12.4-16.6) n=10	moderately broad 2.41 ± 0.08 (2.23-2.49) n=10	spinose 17.9 ± 3.3 (14-26) n=10
spaldingiana (R4-R12)	short 12.2 ± 0.6 (9.8-13.8) n=90	narrow 2.52 ± 0.09 (2.33-2.72) n=90	sparse 9.1 ± 1.8 (6-15) n=90

Subspecies	Uncus Constriction (Min. Width ÷ Max. Basal Width)	Uncus Tip (Max. Tip Width ÷ Max. Basal Width)
retiniana (R2-3)	normal 0.72 ± 0.06 (0.56-0.80) n=13	normal 1.14 ± 0.12 (0.96-1.32) n=13
retiniana(?) (Oregon hybrid) (R1)	normal 0.70 ± 0.06 (0.56-0.77) n=9	normal 1.06 ± 0.07 (0.95-1.17) n=9
spaldingiana (R4-R12)	wide 0.84 ± 0.06 (0.70-0.97) n=78	moderately broad 1.30 ± 0.13 (1.05-1.65) n=78

5

Studies of Pheromone Attraction and Inheritance Using Virgin Females of *Choristoneura* Species, in the Central Sierra Nevada, California

John A. De Benedictis
Department of Entomology
University of California, Davis, California

Andrew M. Liebhold
USDA Forest Service
Morgantown, West Virginia

Jerry A. Powell
Essig Museum of Entomology
University of California, Berkeley, California

Abstract

Pheromone attraction and inheritance of *Choristoneura* species were studied using caged virgin females reared from larvae to attract males in the central Sierra Nevada of California, where adults can be identified as *C. retiniana* or *C. carnana* with certainty. Using pure species, hybrids, and backcrosses, trapping results were consistent with the findings of Liebhold *et al.* (1984), who trapped in a hybrid zone where identities of adults were uncertain. That is, there apparently is simple dominance for production of *C. occidentalis* pheromone over production of *C. retiniana* pheromone which is not sex-linked. If reared from a green or intermediate colored larva, females from a hybrid zone near Sisters, Oregon, attracted only *C. retiniana*. Females reared from brown larvae attracted only *C. carnana*, except one which trapped one moth of each species. There was no evidence that the hybrid zone extends as far north as the type locality of *C. occidentalis*.

INTRODUCTION

Trapping with synthetic sex pheromones has become one of the most important tools for determining relationships among conifer-feeding *Choristoneura* species, but such compounds are only approximations of the actual pheromone. Slight differences between true pheromones and synthetics may alter the biological response of males significantly. Consequently, we supplemented the synthetic chemical baits with caged, live virgin females to provide sources of the true pheromones. Our activities on the west side of the central Sierra Nevada in El Dorado Co., California, demonstrate the utility of caged females in discerning relationships and in investigating pheromone inheritance.

In the central Sierra Nevada, *Choristoneura retiniana* (Walsingham) is dimorphic. About 35% have forewings with a mottling of brown scales on a tawny ground; the remainder have buff forewings with no mottling. The sympatric populations of *C. carnana* (Barnes and Busck) have forewings with broad pale bands transecting the yellow-orange to rust-red ground color. Divergent morphological characteristics (wing length and breadth, basitarsal spination, and uncus shape) also differentiate the species there (De Benedictis Part 4 this vol.).

Until recently, the relationships of the Sierra Nevadan species to populations from other localities were unclear. Powell (1964) inferred relationships based upon distribution, but when biological characters such as larval host preferences (Stehr 1967), immature forms (Harvey and Stehr 1967), and second diapause (Harvey 1967) were used to identify new western species (Freeman 1967), the Californian and other southwestern North American populations were not reconsidered. With the subsequent development of chemical baits based upon female sex pheromones, many populations could be assigned to groups on the basis of chemical attraction.

Beginning in 1978, we tested a series of synthetic pheromone baits for their ability to attract males of various *Choristoneura* populations. The first compound designed to attract *C. retiniana*, the "RET-A" bait (see Daterman *et al.* Part 3 this vol.) for the composition of these lures), attracted moths in southern Oregon but failed to attract any *Choristoneura* in the central Sierra Nevada nor in the mountains of southern California and Nevada during field trials in 1978 and 1979 (Powell and De Benedictis Part 2 this vol.) even though larval coloration and host tree associations suggested a relationship. Larval coloration and preferred host tree also indicated a relationship between *C. carnana* and *C. occidentalis* Freeman, and this was confirmed when the "OXY" bait, a compound that previously had proved attractive to *C. occidentalis* males, drew large numbers of *carnana* males in the Sierra Nevada and elsewhere (ibid.).

We later reared virgin females from field-collected larvae to test attraction of males from one population to females from another, to determine whether the failure of the RET-A bait was a flaw in its formulation, or to ascertain if the Sierra Nevada and southern California mountains populations were unrelated to the northern populations. Females of *C. retiniana* of the northern populations and from the southern California mountains attracted Sierra Nevadan *C. retiniana* and confirmed the suspected relationship (Table 5.1). In concurrent trapping, females of various populations of *C. carnana* attracted sierran *carnana* males, with no interspecific attraction (Table 5.1).

The specific responses of the Sierra Nevada species and the ease of identifying the two species by facies provided a simple bioassay to substantiate the conclusions of Liebhold *et al.* (1984) about pheromone inheritance. After Volney *et al.* (1984) determined that *Choristoneura occidentalis* and *C. retiniana* hybridize in parts of southern Oregon, Liebhold *et al.* (1984) used caged virgin females of both species, their hybrids, and backcrosses to investigate the genetic factors of pheromone production there. Individual moths cannot be identified with certainty at their study sites, because at those localities both species and their hybrids and backcrosses are widely variable with broadly overlapping characteristics (Volney *et al.* 1983, De Benedictis Part 4 this vol.). However, the ranges and distributions of character states in each species vary from those of the other, so Liebhold *et al.* (1984) used a multinomial classification system based upon wing coloration and pattern for comparisons among the trap catches of each female. The array of variation determined whether the catch of any female resembled that drawn to either parental species or differed from both.

Analyses of their trap catches supported the hypothesis of Sanders *et al.* (1977) that an autosomal locus (or loci) controls the specificity of the pheromone blend and that the allele (or alleles) for production of the *C. occidentalis* pheromone is dominant over those producing the *C. retiniana* pheromone. Their analyses could not reveal the precise numbers of either species or hybrids, if any, drawn to each trap and were further complicated because a third species, *C. lambertiana subretiniana* Obraztsov, also is found at their study site and possibly hybridizes with the others. At that time, a synthetic pheromone of *C. lambertiana* had not been developed, and the species was not considered in their studies.

To avoid these complications, we took virgin females of *C. occidentalis* and *C. retiniana*, their hybrids, and backcrosses to the west side of the central Sierra Nevada where the species are distinct, there is no evidence of hybridization, and *C. lambertiana subretiniana* does not occur (Powell and De Benedictis Part 7 this vol.).

Although population densities at our trapping site were much lower than at the southern Oregon sites, the distinctiveness of the species enabled us to determine the exact numbers of each species drawn to females of known genetic lineages. Trap catches could be evaluated without having to rely upon statistical methods to differentiate them and without complications from the presence of hybrids.

We also deployed females reared from larvae collected in a region of hybridization in Oregon. The larvae were categorized by color according to the criterion of Volney *et al.*

(1983). From the results of trapping coupled with the known lineages of females, we could determine whether larval coloration was a reliable indicator of genetic background.

In 1988, we deployed traps with caged females reared from larvae collected in Washington near and well north of the type locality of *Choristneura occidentalis* to delimit the northern boundary of the zone of hybridization of *C. retiniana* and *c. occidentalis.*

MATERIALS AND METHODS

The studies were conducted at or near Blodgett Experimental Forest, El Dorado Co., California. In addition to the preferred larval host of *Choristneura carnana*, Douglas-fir (*Pseudotsuga menziesii* [Mirb.] Franco), and of *C. retiniana*, White Fir (*Abies concolor* [Gord. and Glend.] Lindley), the forest is comprised of several non-hosts, the most abundant of which are California Black Oak (*Quercus kelloggii* Newb.), Pacific Ponderosa Pine (*Pinus ponderosa* Lawson), Incense Cedar (*Calocedrus decurrens* [Torrey] Florin), and Mountain Dogwood (*Cornus nuttallii* Audubon).

Thirty-seven females of *C. retiniana* and *C. carnana* (*sens. lat.*) were used in conjunction with synthetic baits to determine pheromone attraction and selectivity from 1980 to 1982 at Blodgett Forest. Larvae or pupae were collected at several California localities, and reared adult females were caged in traps as described by Powell and De Benedictis (Part 2 this vol.).

Choristoneura retiniana was represented by ten females from the Tehachapi Mountains., Kern Co., one from Modoc Co., and one from Siskiyou Co., California. *C. carnana* females included 3 *C. carnana carnana* from the San Gabriel Mountains, Los Angeles and San Bernardino Cos.; 15 *C. carnana californica* from Lake and Mendocino Cos.; and 9 *C. carnana californica* from the Klamath Ranges of Siskiyou and Trinity Cos., California.

The parental stocks of the virgin females used in tests of pheromone inheritance were non-diapausing laboratory colonies maintained by the USDA Forest Service Pacific Southwest Forest and Range Experiment Station, Berkeley, California. The laboratory colony of *C. occidentalis* originated from Idaho and Montana larval collections (Lyon *et al.* 1972); the *C. retiniana* colony stemmed from the Warner Mts. of northeastern California (J. Robertson, pers. comm.) Hybrids and backcrosses of the laboratory colonies were obtained as described by Liebhold *et al.* (1984). Virgin females were not available for some combinations of hybrids and backcrosses and were only available in limited numbers for most others.

Concurrently, we deployed females reared from larvae collected in a hybridization region near Sisters, Deschutes Co., Oregon. Following the criterion of Volney *et al.* (1983), females were categorized as green, brown, or intermediate according to larval color.

A total of 68 caged females from Sisters and the laboratory colonies were deployed in traps, along with eight empty traps as contols, from July 25-29, 1983, just east of Uncle

Toms Cabin, El Dorado Co., California, at 1,600 m in a mixed forest of similar composition to Blodgett Forest.

During July and August 1988, we trapped with 46 caged females reared from larvae collected in Washington earlier in July on the east side of the Cascade Range. Three were from Satus Pass, Klickitat Co., near the type locality of *Choristoneura occidentalis*, and the remainder from the vicinity of Conconully, Okanogan Co. Forty were deployed in three trap lines at Blodgett Forest from July 19-27, 1988, with the remaining six in another trap line from July 27 to August 6.

In all tests, the females were transported in labeled vials in ice chests to our trapping sites. Each was placed in a small cage to which an identification label was affixed. Cages were either small cardboard boxes with mesh-covered windows (1980-1982) (Powell and De Benedicitis Part 2 this vol.) or aluminum screen cylinders with both ends stapled shut (1983 and 1988). A water-filled vial with a cotton wick was added to each cage. The cages were secured with rubber bands or tape to the inside of Pherocon I-C or II sticky traps (Zoecon Corporation, Palo Alto, California), and moth identification data were written on the outside of the traps. Controls, traps fitted with cages with water vials but without female moths, also were deployed to assure that moths were not attracted to volatile compounds in the sticky substance nor to the caged water vials.

The assembled traps were hung 1.5 to 2 m above the ground in trap lines at intervals of about 8 m. They were placed randomly with regard to tree species, contents of the cage, and the contents of neighboring cages. When retrieved, the numbers and species of trapped males and the condition of the caged females were recorded.

RESULTS

During the trapping tests at Blodgett Forest in 1980-1982, the 27 virgin females of *Choristneura carnana* attracted 209 males of the Sierran population of *C. carnana*, while 62 males of *C. retiniana* were drawn to the 12 caged *retiniana* females. There was no cross-attraction (Table 5.1).

Results of these tests and the subsequent development of effective "RET-C" baits (Daterman *et al.* Part 3 this vol.) determined that in the Sierra Nevada, males of *C. retiniana* were attracted to virgin females of *C. retiniana* from all localities tested and to the RET-C bait, whereas the OXY bait and virgin females of *C. carnana* drew males of Sierra Nevadan *C. carnana*. There was no cross-attraction (Table 2.1 in Powell and De Benedictis Part 2 this vol.). Later tests in northern California indicated that males of *C. carnana* were also attracted to virgin females of *C. occidentalis* from the Forest Service laboratory colony (*ibid.*)

In the 1983 tests, 16 of 51 females of known lineages attracted no moths of either species; the other 35 attracted males of one species only during the four trapping nights (Table 5.2). Among these, 4 females of the *Choristoneura occidentalis* parental stock attracted one *C. carnana* each, while 7 females of the *C. retiniana* parental stock attracted 1 to 25 *C. retiniana* males (65 total). Seven *C. retiniana* X *occidentalis* hybrid females attracted 1 to 9 *C. carnana* males (21 total). The 8 backcrosses to *C. occidentalis* that attracted males drew 1 to 6 *C. carnana* males (23 total), whereas 4 backcrosses to *C. retiniana* drew *C. carnana* males (7 total), and 5 others trapped *C. retiniana* males (44 total).

Ten of the 17 females reared from the larvae collected near Sisters, Oregon, and categorized by larval coloration attracted males (Table 5.3). A total of 110 *C. retiniana* males were drawn to the five females reared from green larvae that attracted moths. Three females reared from brown larvae each attracted one *C. carnana*, while the trap of a fourth contained a male of each species. The female from the intermediate larva attracted 29 *C. retiniana* males.

During July and August 1988, 36 of the 43 females from Okanogan Co., Washington, and 1 of 3 females from Klickitat Co., Washington, attracted male moths. Only *C. carnana* males were attracted (1 to 20 males/trap) (Table 5.4). There was no significant difference between average numbers of males attracted by females collected on *Abies* and those collected on *Pseudotsuga* in Okanogan Co. ($T_{34} = 0.219$, $P = 0.414$).

DISCUSSION

Results of interpopulational trapping with virgin females indicated that males of conifer-feeding *Choristoneura* on the west side of the Sierra Nevada have specific pheromone responses, and the failure of the RET-A lure at most localities (Powell and De Benedictis Part 2 this vol.) was due to its improper formulation rather than differences among *C. retiniana* populations. The attraction of moths to that formulation in Oregon suggests that some males there are less selective and respond to a broader spectrum of the constituent chemicals of the pheromone than do males elsewhere. These non-discriminating males might be the offspring of hybrid matings. If such males exist, they would promote more frequent hybridization there.

The specific responses of males on the west side of the Sierra Nevada and their distinctive appearances provided a reliable bioassay of the pheromones produced by females. Although over 30% (16/51) of the females of known lineages failed to attract any males, results of those that attracted moths are consistent with the hypothesis that an autosomal locus (or loci) controls pheromone production, and the alleles for production of the *C. occidentalis* pheromone are dominant over those for production of the *C. retiniana* pheromone.

The failure of 16 moths to attract any males likely resulted from several factors. Some females may never call, air circulation may have been poor around some traps, and 5 of the 16 females were dead when the traps were retrieved (as were 3 others that attracted moths before dying). It also may be that many of the 16 failed to attract moths because of the low density of *C. carnana* at the trap site. Some of these females may have been sexually active, but no responsive male encountered the pheromone plume.

Such complications and other uncontrollable variables of such a field bioassay require that many more females be tested to provide results that enable better statistical interpretation. However, our trapping results are consistent with and support the simple dominance hypothesis (Sanders *et al.* 1977, Liebhold *et al.* 1984). Under this hypothesis, we would expect that all F_1 hybrids and F_2 backcrosses to *Choristoneura occidentalis* would attract only *C. carnana* and that equal numbers of the F_2 backcrosses to *C. retiniana* would attract only *C. carnana* or only *C. retiniana*. Our results closely matched these expectations (Table 5.2).

In our tests, females of *C. occidentalis*, F_1 hybrids, and backcrosses to *C. occidentalis* all attracted only *C. carnana* males in the Sierra Nevada. Females of *C. retiniana* attracted only *C. retiniana* males. Five of the backcrosses to *C. retiniana* attracted only *C. retiniana*, and four attracted only *C. carnana*. These results were unaffected by.the sex of the females' parents or grandparents, so the alleles for pheromone production apparently are not sex-linked.

The tests of females reared from larvae categorized by color from a natural hybrid zone indicate that green larval coloration is associated with *Choristoneura retiniana*, but that intermediate and possibly brown larval coloration are less reliable indicators of genetic background. This largely conforms with the findings of Volney *et al.* (1983) and Liebhold (1986), who determined that larval coloration is affected by sex and temperature as well as by genetic background. Liebhold (1986) found that all larvae of pure strains of *C. occidentalis* were brown and those of *C. retiniana* were green, as is true for sympatric *C. retiniana* and *C. carnara* in most of California (Powell and De Bendecitis Part 2 this vol.). Larvae of F_1 hybrids were about 2/3 brown with most of the remainder intermediate, larvae of backcrosses to *occidentalis* were more than 95% brown with the remainder intermediate, and about half of those of backcrosses to *C. retiniana* were green with about 2/3 of the remainder intermediate.

Most females reared from brown larvae attracted only *C. carnana* males in our Sierra Nevada trials, but one attracted a moth of each species. It may be that the *C. retiniana* male was not attracted by her pheromone and simply blundered into the sticky trap. However, no other *Choristoneura* male was found in a trap with either the wrong synthetic bait or a caged female of the other species, and none was found in any unbaited control, so it may be that some hybrid females produce an aberrant pheromone that is attractive to either species. Many larvae of F_1 hybrids and F_2 backcrosses are brown (Volney *et al.* 1983, Liebhold 1986). Presumably, some hybrid larvae of other parental lineages are also brown, so this female may have been a natural hybrid with an aberrant pheromone specificity.

If intermediate coloration were indicative of an F_1 hybrids only, females of this classification would be expected to attract only *Choristoneura carnana*. The female from the intermediate larva attracted *C. retiniana*. Liebhold (1986) determined that intermediate larval coloration is most abundant among backcrosses to *C. retiniana*. Such females have a 0.5 probability of attracting *C. retiniana* under the simple dominance model. Presumably intermediate coloration also would be frequent among successive backcrosses to *C. retiniana*, and females of such lineages would have greater than 0.5 probability of attracting *C. retiniana* only.

Our tests did not investigate the inheritance of male response to pheromones, but if a comparable system of inheritance exists, analyses of trap catches in hybrid zones will be further complicated.

Based on trapping results with synthetic pheromones (Powell and De Benedictis Part 2 this vol.), hybridization of *C. occidentalis* and *C. retiniana* apparently does not occur west of the Cascade Range crest. The results of trapping with females reared from larvae collected in Oregon and Washington show that the species hybridize as far north as Sisters, Oregon, but almost certainly do not in the vicinity of Conconully, Washington. We cannot further delimit the hybrid zone because we had insufficient numbers of females from Klickitat Co., Washington, and none from other sites between Sisters and Conconully.

LITERATURE CITED

Daterman, G.E., L.L. Sower, R.E. Stevens, and D.G. Fellin. 1995. Pheromone chemistry and response behavior of conifer-feeding *Choristoneura* in the western United States. Part 3 this volume.

De Benedictis, J.A. 1995. Phenetic studies of spruce budworm populations (*Choristoneura* species). Part 4 this volume.

Freeman, T.N. 1967. On coniferophagous species of *Choristoneura* (Lepidoptera: Tortricidae) in North America. I. Some new forms of *Choristoneura* allied to *C. fumiferana*. Canad. Entomol. 99:449-455.

Harvey, G.T. 1967. On coniferophagous species of *Choristoneura* (Lepidoptera: Tortricidae) in North America. V. Second diapause as a species character. Canad. Entomol. 99: 486-503.

Harvey, G.T., and G. Stehr. 1967. On coniferophagous species of *Choristoneura* (Lepidoptera: Tortricidae) in North America. III. Some characters of immature forms helpful in the identification of species. Canad. Entomol. 99:464-481.

Liebhold, A.M. 1986. Larval coloration of hybrids between *Choristoneura occidentalis* and *C. retiniana* (Lepidoptera: Tortricidae). Canad. Entomol. 118:857-860.

Liebhold, A.M., W.J.A. Volney, and W.E. Waters. 1984. Evaluation of cross-attraction between sympatric *Choristoneura occidentalis* and *C. retiniana* (Lepidoptera: Tortricidae) populations in south-central Oregon. Canad. Entomol. 116:827-840.

Lyon, R.L., C.E. Richmond, J.L. Robertson, and B.A. Lucas. 1972. Rearing diapause and diapause-free western spruce budworm (*Choristoneura occidentalis*) (Lepidoptera: Tortricidae) on artificial diet. Canad. Entomol. 104:417-426.

Powell, J.A. 1964. Biological and taxonomic studies on tortricine moths, with reference to the species in California. Univ. Calif. Publ. Entomol. 32:1-318.

Powell, J.A., and J.A. De Benedictis. 1995. Biological relationships: Host tree preferences and isolation by pheromones among allopatric and sympatric populations of western *Choristoneura*. Part 2 this volume.

Sanders, C.J., G.E. Daterman, and T.J. Ennis. 1977. Sex pheromone responses of *Choristoneura* spp. and their hybrids (Lepidoptera: Tortricidae). Canad. Entomol. 109:1203-1220.

Stehr, G. 1967. On coniferophagous species of *Choristoneura* (Lepidoptera: Tortricidae) in North America. II. Geographic distribution in accordance with forest regions. Canad. Entomol. 99:456-463.

Volney, W.J.A., A.M. Liebhold, and W.E. Waters. 1983. Effects of temperature, sex, and genetic background on coloration of *Choristoneura* spp. (Lepidoptera: Tortricidae) populations in south-central Oregon. Canad. Entomol. 115:1583-1596.

Volney, W.J.A., A.M. Liebhold, and W.E. Waters. 1984. Host associations, phenotypic variation, and mating compatibility of *Choristoneura occidentalis* and *C. retiniana* (Lepidoptera: Tortricidae) populations in south-central Oregon. Canad. Entomol. 116:813-826.

Table 5.1. Moths attracted at Blodgett Forest, El Dorado Co., California, to females reared from field-collected larvae during 1980

Female (Source)	Number of Traps	Number of Males Trapped	
		C. retiniana	*C. carnana*
C. retiniana (Kern Co.)	10	60	0
C. retiniana (Modoc Co.)	1	1	0
C. retiniana (Siskiyou Co.)	1	1	0
Total of all *C. retiniana*	12	62	0
C. carnana carnana (San Gabriel Mts.)	3	0	21
C. c. californica (Lake, Mendocino Cos.)	15	0	65
C. c. californica (Klamath Ranges)	9	0	122
Total of all *C. carnana*	27	0	208

Table 5.2. Trapping results of caged virgin females of known parental lineages

Parental Lineage (Female X Male)[1]	Total Traps	N Traps Containing: Only *C. carnana* (Moths/Trap)	Only *C. retiniana* (Moths/Trap)	Neither, N traps
Lab Colony:				
occ X occ	8	4 (1.00 ± 0.00)	0	4
ret X ret	8	0	7 (10.71 ± 8.56)	1
F_1 Hybrids:				
ret X occ	7	7 (3.00 ± 2.89)	0	0
Backcrosses to *C. occidentalis*:				
occ X (ret X occ)	4	2 (5.50 ± 0.71)	0	2
(ret X occ) X occ	6	5 (2.00 ± 1.22)	0	1
(occ X ret) X occ	2	1 (2.00)	0	1
Backcrosses to *C. retiniana*:				
ret X (ret X occ)	8	3 (1.67 ± 0.58)	3 (8.33 ± 6.66)	2
ret X (occ X ret)	8	1 (2.00)	2 (9.50 ± 12.02)	5

[1] occ = *C. occidentalis*; ret = *C. retiniana*. Parentheses enclose lineage of hybrid parent of F_2 backcrosses.

Table 5.3. Trapping results of caged virgin females reared from larvae collected at Sisters, Oregon, and categorized by color

		Traps Containing:			
Larval Coloration	Total Traps	Only *C. carnana* (Moths/Trap)	Only *C. retiniana* (Moths/Trap)	Both Species	Neither Species
Green	8	0	5 (22.00 ± 25.69)	0	3
Intermediate	1	0	1 (29.00)	0	0
Brown	8	3 (1.00 ± 0.00)	0	1 (1 of each)	4

Table 5.4. Trapping with caged virgin female *C. occidentalis* reared from larvae collected in Washington during 1988

		N Traps Containing		
Source (Host)	Total Traps	Only *C. carnana* (Moths/Trap)	Only *C. retiniana* (Moths/Trap)	Neither Species
Okanogan Co. (*Pseudotsuga*)	28	23 (4.39±4.54)	0	5
Okanogan Co. (*Abies*)	16	13 (4.08±3.30)	0	3
Klickitat Co. (*Pseudotsuga*)	3	1 (2.00)	0	2

6

Foliage-Feeding Lepidoptera of *Abies* and *Pseudotsuga* Associated with *Choristoneura* in California

Jerry A. Powell
Essig Museum of Entomology
University of California, Berkeley, California

John A. De Benedictis
Department of Entomology
University of California, Davis, California

Abstract

During examination of larval hostplant relationships of *Choristoneura*, we worked primarily with low density populations. In such circumstances, associated species of Lepidoptera and sawfly caterpillars often outnumber the *Choristoneura*, and we believe the significance of the associates has been underestimated by many researchers. This report characterizes 40 species of Lepidoptera, provides data on host plant associations, some relative abundance comparisons, and a key for identification of the larvae. Most of the data originate from 240 larval collections that we made during a four-year period. Other records are included from more than 100 collections from prior years in California. In total, we produced 350+ rearing records; about 25% were *Choristoneura*.

About 73% of the species are restricted conifer feeders, 46% specialists on Abietoideae; others are generalists or incidental in mixed forest situations. Two guilds are primarily independent, a more boreal group (10 species) that feeds on *Abies*, and an austral group (7 species) on *Pseudotsuga*. Each of these species feeds mainly on either *Abies* or *Pseudotsuga* and uses both only where the two trees are sympatric, at middle elevations. Five other Lepidoptera are Abietoideae specialists that occur broadly across the ranges of both host tree genera.

The species richness of Lepidoptera feeding on a particular conifer, and of relative abundances, varied widely from place to place and year to year. The three most prevalent associated species were *Dioryctria pseudotsugella* Munroe (Pyralidae), taken in 46 collections at 20 localities; *Coleotechnites* species 2 (Gelechiidae), 34 collections, 26 localities; and *Epinotia radicana* (Walsingham) (Tortricidae), 31 collections, 20 localities. All three are species that range more widely than either *Abies* or *Pseudotsuga* and use both hosts.

INTRODUCTION

A great diversity of lepidopterous larvae feed on the foliage of conifers in western North America, many of them often associated with members of the *Choristoneura fumiferana* complex, the spruce budworms. Furniss and Carolin (1977) list about 60 species of Lepidoptera that occur as specific or general feeders on various Abietoideae in western Nearctic forests, as well as many known exclusively from pines. Stevens *et al.* (1984) report more than 30 species that are associates of *Choristoneura occidentalis* Freeman, principally in Colorado, New Mexico and the Pacific Northwest, about half of which were not mentioned by Furniss and Carolin. Records for many other conifer-feeding species are given by Prentice (1958-1965).

During investigations of the host plant relationships of conifer-feeding *Choristoneura* in California, our target species were *C. carnana carnana* (Barnes and Busck) on *Pseudotsuga macrocarpa* in southern California; *C. carnana californica* Powell, which lives primarily on *P. menziesii* in the Sierra Nevada and Coast Ranges; *C. retiniana retiniana* (Walsingham) in Siskiyou and Modoc Cos. and *C. retiniana spaldingiana* Obraztsov in the Sierra Nevada and Tehachapi Range, both on *Abies*; and *C. lambertiana subretiniana* Obraztsov in the Sierra Nevada, which feeds primarily on *Pinus* but is found occasionally on *Abies*. We worked primarily with low density populations of the budworm complex (Powell and De Benedictis Part 2 this vol.). In such circumstances, associated species often approach or outnumber individuals of *Choristoneura*, making discovery and monitoring of the *Choristoneura* populations difficult.

We believe that the significance of associated species of Lepidoptera and Hymenoptera (Pamphiliidae, Tenthredinidae) caterpillars has been underestimated by many *Choristoneura* researchers. In the field, younger instars of the moth larvae frequently are extremely difficult to differentiate from those of the budworms, particularly *C. retiniana* (Walsingham). This report briefly characterizes 40 species that we encountered in collections from *Abies* and *Pseudotsuga* (Table 6.1). We collected fewer than half of the 33 species mentioned by Stevens *et al.* (1984). Several of the associated species that are known to occur in California, but were not detected during our larval surveys, are included in our diagnostic key to larvae (Appendix 6.2).

METHODS

Surveys for larvae of *Choristoneura* were carried out intermittently from 1970 to 1982. In 1970 we examined *Abies* in southern Oregon, and in California in Modoc Co. and on the east side of the northern Sierra Nevada. In June 1974 we surveyed *Abies* in the Mt. Shasta and Manzanita Mountain, Modoc Co., areas, and the U.S. Forest Service intensively monitored outbreak populations of *C. retiniana* on *Abies concolor* in the Modoc region. Preliminary samples taken in May and June, and later collections prior to and following a spray project, were sent to Berkeley for rearing (Powell and Donahue 1974).

In 1979 we began a more comprehensive sampling program, concentrating on *Pseudotsuga* as well as *Abies*, in various parts of California (Powell and De Benedictis Part 2 this vol.). During 1979-1982 we processed approximately 240 larval collections from these two hosts.

Most of the collections were made by searching the lower canopy, 1-3 m high. The 1974 USFS samples consisted of four 15-inch branch terminals taken by pole pruner from mid-crown (ca. 5-6 m high) from each of 30 trees per site. In 1979 we tried to sample more thoroughly than in other years, particularly *Pseudotsuga*. We employed visual search, beating-sheet samples, and random clip samples by pole pruner at 4 to 6 m heights at each locality. We collected Noctuidae and Geometridae by beating sheet more effectively than by visual searches, so sites surveyed in 1979 are better represented for those taxa than in later years when we seldom used beating sheets. All *Choristoneura* populations sampled in 1979 were low density (0-6 larvae/search hr), and no budworms were taken in pole pruner collections. Hence, we did not continue to use pole pruners for random survey in subsequent seasons.

Larvae were placed in translucent plastic bags and transported in field iceboxes. Collections in the 1974 Modoc survey were processed as follows: larvae were sorted in a temporary lab at Alturas, California, placed in plastic petri dishes with synthetic diet, and forwarded to Berkeley; usually there were 7-10 days between collection and receipt; samples were taken 24 hrs prior to and 6 and 16 days after a spray treatment on 6 of 9 sites sampled. From the pre-spray samples, about 100 *Choristoneura retiniana* and all associated larvae judged to be other species were forwarded. The total number of larvae processed was not revealed to us, so the percentage of associated species cannot be calculated. From the post-spray samples, all available *C. retiniana* as well as associated species were shipped.

In the lab, the 1974 larvae were housed in plastic petri dishes and reared on synthetic diet (Lyon *et al.* 1972). In other years larvae were placed in plastic bags with paper toweling and twigs of host foliage. When Noctuidae and Geometridae matured, they were moved to plastic boxes with 1-2 cm of sterile sand as a medium for pupation.

Collection data and rearing records, preserved larvae, reared adult specimens of moths and parasitoids, photographs, pupal shells, etc., were associated by rearing lot numbers. Each collection was assigned a number corresponding to the date collected (e.g., 70F25 = 1970, June, 25th collection), and this was given to all associated data and specimens. The records are summarized in a database program.

RESULTS

We processed more than 350 collections of lepidopterous larvae from *Abies* and *Pseudotsuga* during 1970-1982. In addition to *Choristoneura* (90 rearing records, summarized by Powell and De Benedictis Part 2 this vol.), we obtained 46 species of moths, represented by more than 250 rearing records. Several of these were incidental occurrences of non-conifer- feeding species, scavengers, and/or single individuals of uncertain taxonomic status or biological significance. Table 6.1 summarizes records for the remaining 40 species: *Choristoneura retiniana, C. carnana*, and 38 other species. The data are given in Appendix 6.1.

DISCUSSION

Host Preferences

Numbers of collections from *Abies* and *Pseudotsuga* (Table 6.1) reflect concentration of our sampling effort in particular geographical and ecological areas. We did not attempt to sample randomly in all parts of California's coniferous forests, nor equally on each host species. The designation of host preference in Table 6.1 takes into account records from other regions of western North America (Prentice 1958-1965, Furniss and Carolin 1977, Stevens *et al.* 1984).

Among the 38 associated species that we reared in California, most are restricted conifer feeders. About 17 (46%) are specialists on Abietoideae; another 10 species (27%) are generalists on conifers, including 3 that are primarily pine feeders (*Choristoneura lambertiana subretiniana, Chionodes sabinianae, Zelleria haimbachi*). Nine species are general feeders on hardwoods or other plants and probably only occasionally feed on conifers: *Hyalophora euryalus*, *Egira crucialis, E. hiemalis*, and 6 tortricines: *Archips argyrospila*, *Argyrotaenia citrana*, *Choristoneura rosaceana*, *Clepsis persicana forbesi*, *Amorbia cuneana*, and *Sparganothis senecionana*. Finally, 2 species, *Telphusa sedulitella* and *Ypsolopha cervella* are oak feeders, but their larvae wander, and in mixed oak-conifer stands late instar larvae occur on Douglas-fir often enough to warrant discussion.

In California, Lepidoptera feeding on Abietoideae generally comprise two somewhat independent guilds, a boreal group that feeds on *Abies concolor* and *A. magnifica*, and a more austral set on *Pseudotsuga menziesii* and *P. macrocarpa*. Thus, their distributions roughly correspond to those of *Choristoneura retiniana* and *C. carnana*, respectively (Powell and De Benedictis Part 2 this vol.). These guilds are not mutually exclusive, but most members are associated with either *Abies* or *Pseudotsuga* and feed on both only where the two trees are sympatric. Most of these species may use both hosts in the relatively narrow (ca. 500 m elevation) zone of sympatry of *Abies* and *Pseudotsuga* in California, but we have not sampled sufficiently in this ecotone to document host preferences in mixed stands.

There are 10 associates of *Choristoneura retiniana* on *Abies* that occur primarily at higher elevations and not in lower areas with *Pseudotsuga* in absence of *Abies*: *Syngrapha celsa*, *Epinotia meritana*, *E. trossulana*, *Archepandemis coniferana*, *Argyrotaenia dorsalana*, *A. lautana*, *A. provana*, *Clepsis persicana*, *Chionodes abella*, and *Coleotechnites* species 1. Conversely, the associates of *Choristoneura carnana* at lower elevations on *Pseudotsuga* are: *Lophocampa argentata*, *Feralia deceptiva*, *Neoalcis californiaria*, *Cochisea sonomensis*, *Gabriola dyari*, *Nepytia umbrosaria*, and *Coleotechnites* species 3.

There are five species of Abietoideae-feeding specialists that occur across the ranges of both *Abies* and *Pseudotsuga* and could be classified as members of both the boreal and austral guilds: *Enypia packardata*, *Dioryctria pseudotsugella*, *Epinotia radicana*, *Acleris gloverana*, and *Coleotechnites* species 2.

Species Richness and Relative Frequency of Collection

The numbers of Lepidoptera species (richness) feeding on a particular conifer and the numbers of individual larvae of each (relative abundance) varied widely from place to place. Overall, we produced 350+ rearing records, the emergence of one or more adults of a species from one host species collected at one place and time. Only about 25% (n=90) of these were *Choristoneura*, even though our efforts were directed to their collection. This was because we surveyed primarily in low density populations of *Choristoneura* (fewer than 10 larvae/search hour), and larvae of associates frequently were more abundant. By contrast, most previous studies of *Choristoneura* have concentrated on outbreaks, where budworms cause noticeable defoliation and vastly outnumber the associated species.

Our survey yielded *C. retiniana* in 52 samples from 28 localities, while *C. carnana* was reared in 32 collections from 17 localities (Powell and De Benedictis Part 2 this vol.). Three of the associated species were similarly well represented: *Dioryctria pseudotsugella* (46 collections, 20 localities), *Coleotechnites* species 2 (34 collections, 26 localities), and *Epinotia radicana* (31 collections, 20 localities). Not surprisingly these are species that range more widely than either *Abies* or *Pseudotsuga* and use both hosts. Four additional species were reared from 10 or 11 sites: *Argyrotaenia dorsalana* (18 collections), *Chionodes abella* (15), *Acleris gloverana* (13), and *A. provana* (11). All of the latter group are primarily associated with *Abies* in California. The first 5 of these 7 prevalent associated species are listed among the 6 major associates of *C. occidentalis* and *C. retiniana* in eastern Oregon and Washington (Carolin 1980), and 4 of the 7 were identified as components of an outbreak of defoliators in Idaho (Markin 1982).

Even though *Choristoneura* are the dominant herbivores during outbreaks ascribed to them, there may be appreciable proportions of other Lepidoptera in the guild. Carolin (1980) monitored the relative abundance of associated species in several populations of high density *C. occidentalis* or *C. retiniana* in Oregon. In general, he found similar trends in abundance from year to year at a particular site; as numbers of budworm rose, so did those of associated Lepidoptera. However, the proportion of associates varied dramatically from year to year and from site to site within one year. For example, in sprayed plots the proportion of species other than budworm rose from 88% in 1953 to 100% in 1954, then dropped to 36% the following year at Mt. Hood, while reverse trends occurred at other sites in the same years, dropping from 60% to 14% and climbing to 88% at Eugene, and from 27% to 0 to 65% at Umatilla.

In unsprayed areas, Carolin found the proportion of species associated with high density *Choristoneura* populations (150-2,000 larvae/100 15-inch twigs) to be generally low, but there were parallel trends of increase from 1953 to 1958 at several *C. occidentalis* sites. In these circumstances 25% to 51% of the sampled larvae at four localities were caterpillars other than budworms.

In Idaho, *C. occidentalis* comprised 48% of the defoliators in 1976, with five species of Tortricidae, Pyralidae and Geometridae making up most of the remainder (Markin 1982).

During the 1974 monitoring of the *C. retiniana* outbreak and effects of a spray program in the Modoc-Lassen area, we processed more than 7,000 larvae, of which 33% were *Dioryctria* (Powell and Donahue 1974). In post-spray samples, which included all available larvae of C. retiniana as well as those of associated species, *D. pseudotsugella* comprised 29% of 6-day and 21% of 16-day samples from sprayed plots, and 15% and 7% of corresponding samples in unsprayed plots. In this instance, 10 other species of Lepidoptera were insignificant, making up 0.5-2.0% of the samples. We did not find *Choristoneura*, nor *Dioryctria*, nearly so numerous at other sites in California, but at Tehachapi Mountain, where *C. retiniana* was moderately abundant (40-50 larvae/search hr) in 1980, 1981 and 1982, *Dioryctria* was more numerous than at other sites, comprising about 20% of the sample in 1980. However, all associates including *Dioryctria* made up only 13% of our 3-year collections at Tehachapi. In 1989, this caterpillar community had severely declined, with < 6 *C. retiniana*/search hr, no *Dioryctria* discovered, and other associates negligible.

Only rough estimates of relative abundances can be inferred from the frequencies of species reared in our survey, because we sought to maximize detection of species, did not sample randomly, and usually had small sample sizes (Table 6.2). In populations other than Tehachapi, associated species ranged from 20% to 76% of the larvae, when samples were pooled by regions (Table 6.2). Overall, about 42% of all specimens reared during 1979-1982 were non-*Choristoneura*. In sparse populations, there were widely varying proportions of budworm and caterpillars of other species. Some examples are as follows.

In Sonoma County, all larvae obtained from *Pseudotsuga* during a 6-hr search in 1980 in the vicinity of a 1967 light trap collection of *C. carnana californica* proved to be other species, primarily three tortricids (*Epinotia radicana*, *Acleris gloverana*, and *Archips argyrospila*).

At Cobb, Lake Co., where we collected *Choristoneura carnana* larvae at rates of 3-10 larvae/search hr, samples taken in 1979-80 contained 74% *C. carnana*, with the remaining 26% made up by 8 other species. *Pseudotsuga* at Blodgett Forest, El Dorado Co., with lower larval density (1-2 larvae/search hr) produced 57% *Choristoneura.*

By contrast, *Abies* consistently yielded low proportions of *C. retiniana*, except at Tehachapi, and only 6.6% of moths from our *A. concolor* samples at Blodgett Forest during two seasons were *C. retiniana*. In a collection at Hockaday Springs, Siskiyou Co., only one of 20 green larvae feeding on *A. concolor* proved to be *C. retiniana*. The remainder included three tortricines, *Argyrotaenia dorsalana*, *Clepsis persicana forbesi*, and *Sparganothis senecionana*, with the last, a general feeder, most common.

In the Lake Tahoe area several collections produced an aggregate of 30% *C. retiniana*, along with 8 other species. Similarly, in the Manzanita Mtn. area near Adin, Modoc Co., with *C. retiniana* at low population levels, 1980 collections yielded only 27% *C. retiniana* (contrasted with 67% in 1974), 33% *Epinotia radicana*, and 40% among 5 other species.

TAXONOMIC AND BIOLOGICAL SYNOPSES OF THE REARED SPECIES

Gelechiidae

Coleotechnites apicitripunctella complex

Coleotechnites apicitripunctella (Clemens, 1860) is an eastern Nearctic species that feeds on *Tsuga* (Prentice 1965). About 30 related species have been described from conifers, including *Cupressus*, *Juniperus*, *Thuja*, *Pinus*, and various Abietoideae. A comprehensive taxonomic assessment is needed to verify the degree of host specificity and establish whether all the names represent valid species. None of the Abietoideae-feeders is described from the western United States, so it is likely that one or more of the species we encountered is undescribed.

The larvae of the species we observed mine the needles of *Abies* or *Pseudotsuga*. They enter the needle, usually from the underside, leaving a circular hole, through which frass is ejected. After a needle is about half mined, the larva moves to an adjacent one. In the process several are mined and are held together by a characteristic sheet of silk. The empty mines are numerous and conspicuously paler than surrounding foliage. Although the larvae are distinguishable from young *Choristoneura* on close examination, in the field the mining and webbing habits are very similar to those of early instars of *C. retiniana* on *Abies*.

Coleotechnites species 1

This species superfically resembles *Coleotechnites* (*Eucordylea*) *atrupictella* (Dietz). We reared adults only from *Abies concolor* in the Warner Mountains., Modoc Co. Larvae of both this and *Coleotechnites* species 2 were detected at five Modoc sites on *A. concolor*, but they were not discovered together. Three of four collections in 1970 were species 2, while all five collections in 1974 were species 1, and two collections in 1980 were species 2.

In Canada, populations referred to *C. atrupictella* occur from Newfoundland to British Columbia and feed on various Abietoideae; 40% of 400 collections came from *Picea*, 32% from *Pseudotsuga*, and 10% from *Abies* (Prentice 1965).

Coleotechnites species 2

This was one of the most ubiquitous non-*Choristoneura* encountered during our survey. Larvae occurred, often abundantly, on *Abies concolor* in the Salmon Mountains-Mt. Shasta area, Modoc region, Sierra Nevada, Tehachapi Range, Mt. Pinos and San Gabriel Mountains, and on *A. magnifica* at Lake Tahoe. They were less prevalent but generally distributed on *Pseudotsuga menziesii*, from the Mt. Shasta region, southward in the Coast Ranges and Sierra Nevada. We found larvae mining in both *Abies* and *Pseudotsuga* at Prather Creek north of Mt. Shasta and at Blodgett Forest, El Dorado Co.

Adults reared from both hosts vary considerably in forewing color and pattern, but there are no consistent differences between *Pseudotsuga* and *Abies* moths. Phenotypically, *Coleotechnites* species 2 is similar to *C.* (*Evagora*) *piceaella* (Kearfott), a spruce-feeder in eastern North America that ranges west to Alberta (Prentice 1965).

Coleotechnites (*Pulacavaria*) *granti* (Freeman), which was described from British Columbia on *Abies grandis*, may be the nearest relative of species 2. *A. grandis*, a coastal species in California, was not surveyed during our study. Comparison of *Coleotechnites* sp. 2 adults with paratypes of *C. granti* showed phenotypic differences distinguishing the two.

Coleotechnites species 3

Probably this is an undescribed species. It occurred exclusively on *Pseudotsuga macrocarpa* in the San Gabriel and San Bernardino Mountains. The adults are larger than those of species 2, and the forewing lacks the ochreous or tan hue in the pale areas; the pattern resembles some pine-feeding *Coleotechnites*, such as *C.* (*Evagora*) *resinosae* (Freeman).

Telphusa sedulitella (Busck)

Gelechia sedulitella Busck, 1910, Proc. Entomol. Soc. Wash. 11:181 [1909].
Telphusa sedulitella; Keifer, 1933, Calif. Dept. Agric. Mo. Bull. 22:359.

This species is an abundant member of the spring caterpillar community on new foliage of oaks, especially *Quercus agrifolia*, in California. We reared *T. sedulitella* in Marin and Napa Cos. from *Pseudotsuga menziesii* that was growing in mixed oak-conifer stands. We assume that the larvae were incidental, late instar migrants from oak. Stevens *et al.* (1984) illustrate the adult of this species and describe its development on *Abies grandis*. Their description of the larva, however, is that of another species. Larvae of *T. sedulitella* are entirely green, including the head, without longitudinal stripes, and they become tinged with pink dorsally at maturity. They have conspicuous black setae on the anal shield (Keifer 1933, UCB specimens).

Braun (1921) also briefly characterized a larva supposed to be *T. sedulitella* (=*agrifolia* Braun), as having dark plum-colored head and thoracic segments 1-2, the remainder of the body grayish with dark pinacula. She gave no indication as to the source of the note on the larva or its association with reared adults. This diagnosis could apply to *Chionodes braunella* (Keifer), a legume-feeder, but there is no such larva feeding on *Q. agrifolia.*

Prentice (1965) lists *T. sedulitella* from Vancouver Island, with collections in several years from *Q. garryana*, and three collections from *Salix*, while Clarke (1934) reported a larval collection from *Corylus*. We have processed more than 12,000 larval collections at Berkeley during the past 35 years, the majority of them external-feeding Microlepidoptera, and we have not found *T. sedulitella* to be a general feeder. We have reared *T. sedulitella* from more than 50 collections on several species of *Quercus*, and single individuals twice from *Arctostaphylos* (Ericaceae) growing with oaks, but not from any other hostplant, other than the two records from Douglas-fir.

Chionodes abella (Busck)

Gelechia abella Busck, 1903, Proc. U.S. Natl. Mus. 25:889.
Chionodes abella; Busck, 1939, Proc. U.S. Natl. Mus. 86:574.

The typical form of this moth, from Colorado and Utah, has black forewings with the apical one-third white. It was reported as a rare associate of *Choristoneura occidentalis* in the southern Rocky Mountains by Stevens *et al.* (1984), who did not list a host tree. California populations referred to this species lack the apical white area of the forewing, having instead whitish transverse bands. Some individuals are nearly completely blackish, but all have consistent genitalia characters (Hodges pers. comm.).

We collected larvae of *Chionodes abella* throughout the sampled range of *Abies concolor* in California from Tehachapi northward, and on *Pseudotsuga menziesii* at Blodgett Forest, El Dorado Co., where *C. abella* also fed on *A. concolor*. We did not find larvae in southern California, nor in the North Coast Range, where *C. abella* was replaced by *C. periculella* on *Pseudotsuga.*

The larvae feed in the newly developing foliage in June, along with *Choristoneura retiniana*. They are reddish, thus superficially resembling *Choristoneura carnana* or *Dioryctria* in early instars. Mature larvae of *Chionodes abella* are slender, reddish brown, with a mid-dorsal line and three parallel, dorsolateral and lateral whitish stripes that are incomplete, irregularly interrupted, and broken in the intersegmental areas. The head and thoracic shield are amber brown.

Stevens *et al.* (1984) state that there are two undescribed species of this complex associated with *Choristoneura occidentalis*, one in southern Oregon, the other more widespread in the northwestern states. By contrast, Prentice (1965) does not record any *Chionodes* in Canada on *Abies* or *Pseudotsuga.*

Chionodes periculella (Busck)

Gelechia periculella Busck, 1910, Proc. Entomol. Soc. Wash. 11:178 [1909].
Chionodes periculella; Busck, 1939, Proc. U.S. Natl. Mus. 86:574.

This gelechiid is closely related to *Chionodes abella*, but the forewings are nearly uniformly dark gray, with the transverse whitish markings reduced and not as variable. *C. periculella* has remained poorly known subsequent to its original description from Humboldt Co., California. We reared four specimens at two sites in Napa Co. and one believed to be this species from Lake Co., from *Pseudotsuga menziesii*, and it was reared from Douglas-fir at Ashland, Oregon (NMNH). Evidently nothing has been reported previously on its biology. The larva is dark reddish brown marked with longitudinal pale stripes. Probably it is similar to that of *C. abella*, but none of the *C. periculella* was preserved.

Chionodes sabinianae Powell

Chionodes sabinianae Powell, 1959, Entomol. News 70:127.

This is a Californian endemic that feeds primarily on pines. The original collections were reared from larvae feeding in staminate cones of Digger Pine, *Pinus sabiniana*, and larvae subsequently have been collected from *P. coulteri* (JAP 76D48). We reared *C. sabinianae* from *Abies concolor* at Tehachapi Mountain, where it fed with *C. abella*. Digger pine grows nearly to the elevation of *Abies* in the region, so the host switch to *Abies* is not terribly surprising.

The mature larva of *C. sabinianae* is primarily whitish with reduced reddish markings that form parallel dorsolateral and lateral bands the length of the body and irregular blotches on each segment. The head is orange-brown blotched with brown laterally and posteriorly; the prothoracic shield is dark brown (Burdick and Powell 1960).

Plutellidae

Ypsolopha cervella (Walsingham)

Cerostoma cervella Walsingham, 1881, Proc. Zool. Soc. London 1881:307
Ypsolopha cervella; Heppner and Duckworth, 1983, *in*: Hodges (ed.), Check List Lepid. Am. North of Mexico, p. 26

Ypsolopha cervella is another abundant member of the caterpillar community on various species of *Quercus* (Fagaceae) in California, and it ranges north to British Columbia. The larvae typically wander from oaks at maturity and form cocoons on nearby objects, including other trees. As a result, in mixed oak-conifer woods, *Y. cervella* sometimes occurs with *Choristoneura* on Douglas-fir. We reared three individuals in Napa Co. and one in Siskiyou Co. from larvae on *Pseudotsuga menziesii,* although probably they did not feed there. Stevens *et al.* (1984) also reported this species on Douglas-fir in British Columbia.

Ypsolopha nella (Busck) is a common associate of *Choristoneura occidentalis* on *Abies concolor*, evidently its native host, in the Rocky Mountains of Colorado and New Mexico (Stevens *et al.* 1984). Adults of the two *Ypsolopha* species are similar in forewing color pattern and variation. *Y. cervella* is larger, has broader forewings that have a brassy, shiny appearance; *Y. nella* has a dull, slightly mauve tint.

The larva of *Y. cervella* is bright green including the head; the abdominal prolegs are elongate, and the pinacula are conspicuously black. The larval color in *Y. nella* is yellowish with brownish dorsal banding but is variable according to Stevens *et al.* The black pinacula are distinctive.

Other species of *Ypsolopha* are found as occasional associates of conifer-feeding *Choristoneura*, presumably owing to the wandering habits of late instar larvae. We reared *Y. dentiferella* (Walsingham), which feeds on Caprifoliaceae, from a cocoon found on *Abies magnifica* at Lake Tahoe.

Yponomeutidae

Zelleria haimbachi Busck

Zelleria haimbachi Busck, 1915, Proc. Entomol. Soc. Wash. 17:91.

This species, commonly known as the Pine-needle Sheathminer, is transcontinental in distribution. The larvae are recorded from several species of pines (Stevens 1959, Prentice 1965, Furniss and Carolin 1977). In California, *Pinus ponderosa* and *P. jeffreyi* were regarded as the primary hosts of *Z. haimbachi* by Stevens (1959), although many pine species were used in a nursery situation. Stevens also stated that there are isolated records from Canada of its occurrence on *Picea* and *Abies balsamea*, but Prentice (1965) listed only *Pinus* among 350+ rearing records across Canada.

We found the larvae on *Pseudotsuga menziesii* (n=2) in the Trinity Mountains and on *P. macrocarpa* (n=2) in the San Gabriel Mountains, as well as on *Pinus jeffreyi* in the Sierra Nevada. The first instar is a needle-miner and spends the winter in that stage. In spring the larvae emerge from the mines and on pine feed at the base of the needle clusters, within the sheaths, moving to several new sheaths during development. Mature larvae are slender, tan with broad, dull orange lateral stripes along the length of the body (Stevens 1959).

Tortricidae: Eucosmini

Zeiraphera pacifica Freeman

Zeiraphera pacifica Freeman, 1966, J. Res. Lepid. 5:588.

We did not encounter this species during larval surveys in California, but it was reared from *Abies concolor* by Volney *et al.* at four of their study sites in southern Oregon (UCB specimens). Adults of *Z. pacifica* consistently were taken at blacklight at Lake Tahoe by N. Westerland (LACM), but this species has not been recorded at lower elevations in California. These records suggest that the species is limited to *Abies* in this part of its range.

According to Stevens *et al.* (1984), sympatric populations of *Z. pacifica* and *Z. hesperiana* Mutuura and Freeman in Idaho tend to be host specific, the latter on Douglas-fir and *Z. pacifica* on *Abies*. The larvae are tan or brownish, darker dorsally, with an orange head and tan-orange thoracic shield that is conspicuously bordered with black posteriorly.

Epinotia radicana (Heinrich)

Griselda radicana; Heinrich (not Walsingham, 1879), 1923, U.S. Natl. Mus. Bull. 123:186.
Griselda radicana Heinrich; Obraztsov, 1964, Bull. Zool. Nomencl. 21:144.
Epinotia radicana; Brown, 1983, Entomography 2:105.

Nomenclatural confusion was created when Heinrich misidentified the species described as *radicana* by Walsingham, which is now considered to be a species of *Epiblema*, when Heinrich described the genus *Griselda*. Obraztsov (1964, 1965) stabilized the names by requesting the International Commission on Zoological Nomenclature to fix *Griselda radicana* Heinrich as the valid name for the widespread conifer-feeding species. Subsequently R. L. Brown treated *Griselda* as a subjective synonym of *Epinotia*.

Epinotia radicana appeared in more than 2,000 collections of the Forest Lepidoptera of Canada survey (Prentice 1965), on various conifers, including four records from *Thuja* and *Juniperus* (Cupressaceae). However, nearly all the collections were from *Picea* (60%), *Pseudotsuga* (22%), or *Abies* (16%).

This was the most commonly detected non-target tortricid during our research, occurring in 9% of all collections, from 20 localities. It fed on *Abies* in the Mt. Shasta-Warner Mountains area and in the Sierra Nevada south to El Dorado Co., and on *Pseudotsuga* at Mt. Ashland, in the Coast Range south to Napa Co., and in the central Sierra Nevada, using both hosts at Blodgett Forest. *E. radicana* has not been recorded in the Coast Ranges south of San Francisco Bay nor in southern California.

The larvae are tan with broad, diffuse rust-orange bands, similar in color to early instar larvae of *C. carnana*. The two are sympatric, but larvae of *E. radicana* tend to complete development late in the season, with adults emerging in August and September, after those of sympatric *Choristoneura*.

Epinotia meritana Heinrich

Epinotia meritana Heinrich, 1923, U.S. Natl. Mus. Bull. 123:226.

This species, known as the White Fir Needleminer, was described from Utah and is widespread in the western United States. Outbreaks causing defoliation of Red Fir and White Fir have occurred periodically in Utah, Arizona and the Sierra Nevada in California (Furniss and Carolin 1977). We encountered only occasional individuals, on *Abies concolor* in the Warner Mountains and Tehachapi Range. The larvae, which live as needleminers, are cream-colored to yellowish green with dark head and thoracic shield and can be confused with the mining instars of *Choristoneura retiniana*.

Epinotia trossulana (Walsingham)

Grapholitha trossulana Walsingham, 1879, Illus. Lepid. Het. Brit. Mus. 4:67.
Epinotia trossulana; Heinrich, 1923, U. S. Natl. Mus. Bull. 123:231.

Larvae of *Epinotia trossulana* mine within needles or bundles of needles of *Abies*. The species is generally distributed in California in association with *A. concolor*, but we reared it only in the northern counties and from *A. magnifica* at Lake Tahoe. Neither this nor the preceding species was recorded in Forest Lepidoptera of Canada (Prentice 1965) nor among associates of budworms compiled by Stevens *et al.* (1984).

Tortricidae: Tortricini

Acleris gloverana (Walsingham)

Lophoderus gloveranus Walsingham, 1879, Illus. Lepid. Het. Brit. Mus. 4:14.
Acleris gloverana; Powell, 1962, Canad. Entomol. 94:836.

Acleris gloverana was described originally from Mt. Shasta, California, but its taxonomic status was not recognized for nearly a century. As a result, much of the early literature on this insect is recorded under the name *Acleris* (=*Peronea*) *variana* (Fernald 1886) (e.g., Prentice 1965). Females of the latter, an eastern species, possess large tufts of abdominal scales, which are deployed covering the eggs, an adaptation that is lacking in *A. gloverana* (Powell 1962). If these and other morphological differences are considered to be trivial, and the whole complex is treated as a single species, *A. gloverana* is the valid name on the basis of priority.

The Western Black-headed Budworm is widespread in western North America and undergoes outbreak population levels causing defoliation of various Abietoideae (Furniss and Carolin 1977, Stevens *et al.* 1984). In California, *A. gloverana* has been collected from *Tsuga heterophylla*, but it is primarily associated with *Abies concolor*. It was the most widely encountered tortricine other than *Choristoneura* in our collections. Larvae occurred on *Abies* from the Warner Mountains to the central Sierra Nevada, and on *Pseudotsuga* in the Trinity region, North Coast Range, and Sierra Nevada; they fed on both hosts at Blodgett Forest, El Dorado Co.

The larvae are green with a black head and prothoracic shield until the final instar, which has a brown head and weakly brownish shield. Early instars superficially resemble those of several other tortricids including *Choristoneura retiniana*. The adults, which are polymorphic in forewing color pattern, emerge late in the season, usually after sympatric *Choristoneura*.

Tortricidae: Archipini

Argyrotaenia provana (Kearfott)

Olethreutes provana Kearfott, 1907, Trans. Am. Entomol. Soc. 33:16.
Argyrotaenia provana; Freeman, 1958, Canad. Entomol. 90, Suppl. 7:54.

Argyrotaenia provana was described from Vancouver Island, and it ranges along the coast to Sonoma Co., California, as well as widely inland to Colorado, New Mexico, and eastern Arizona. The larvae feed on *Pseudotsuga menziesii, Abies concolor,* and *A. magnifica* (Furniss and Carolin 1977, and present data). We detected larvae only in small numbers (n=1-3) in the Trinity and Warner Mountains and at Donner Lake in the central Sierra Nevada. Adults emerged late in the season, from the end of July to September. According to Stevens *et al.* (1984), the larva resembles that of *A. dorsalana* but has dorsolateral and lateral, yellowish stripes that are lost in the final instar.

Argyrotaenia citrana (Fernald)

Tortrix citrana Fernald, 1889, Entomol. Am. 5:18.
Argyrotaenia citrana; Basinger, 1935, Calif. Dept. Agric. Bull. 24:233.

Members of the *Argyrotaenia franciscana-citrana* complex are polyphagous; larvae have been recorded from an immense diversity of plants, including native *Cupressus* (Cupressaceae), *Sequoia* (Taxodiaceae), and *Pinus, Tsuga*, and *Picea* in urban situations (De Benedictis *et al.* 1990, Powell 1964 and unpubl. data). Populations are limited to coastal and low elevation areas, so generally do not inhabit conifers with *Choristoneura*. However, *A. citrana* and *C. carnana* occur sympatrically in coastal Sonoma and Mendocino Cos. and in the Santa Cruz Mountains.

We found larvae on *Pseudotsuga* near Occidental, Sonoma Co., in a native situation. The larvae are pale to dull green, usually showing conspicuous dorsolateral, pale stripes, and therefore are easily distinguished from *C. carnana* caterpillars. None of the other conifer-feeding species of *Argyrotaenia* is sympatric with *A. citrana*.

Argyrotaenia dorsalana (Dyar)

Tortrix dorsalana Dyar, 1903, Proc. Entomol. Soc. Wash. 5:231.
Argyrotaenia dorsalana; Freeman, 1958, Canad. Entomol. 90, Suppl. 7:55.

This species was originally described from specimens that were labeled as having been reared from oak at Williams, Arizona. All subsequent biological records indicate that *A. dorsalana* is a conifer-feeder (Powell 1964, Prentice 1965, Furniss and Carolin 1977, Stevens *et al.* 1984). The typical montane form, which has yellow forewings with variable, sexually dimorphic, brown markings, has been reared from various Abietoideae. In British Columbia, 4 of 17 single-specimen collections were from *Thuja* (Cupressaceae) (Prentice 1965). Volney *et al.* reared *A. dorsalana* in numbers from *Abies concolor* at Mt. Ashland, and at three sites in Klamath County, Oregon, from *A. magnifica*, *Pseudotsuga menziesii*, and *Tsuga mertensiana* (UCB specimens).

We found larvae of *A. dorsalana* on *Abies* in the Salmon Mountains, Warner Mountains, and at Lake Tahoe. We did not detect it on *Pseudotsuga*, although this is a preferred host in British Columbia (Prentice 1965). Collection records indicate that this insect is more boreal than is *Pseudotsuga* in most of California. The larvae are bright green with a slightly tan head and thoracic shield and are not superficially distinguishable from those of *Choristoneura retiniana* except by size.

Populations in the Great Basin have rust-colored forewings, and there are intermediate specimens from Utah (Powell 1964 and unpubl. data). Circumstantial evidence, including laboratory rearing from eggs (JAP 77G6), indicates that pines are the hosts in these areas.

Argyrotaenia lautana Powell

Argyrotaenia lautana Powell, 1960, Pan-Pac. Entomol. 30:90.

This species occurs in the Greenhorn Mountains, Tehachapi, and Transverse Ranges of southern California, and adults sometimes are fairly abundant at blacklight. We did not find larvae numerous but reared *A. lautana* with *Choristoneura retiniana* from *Abies concolor* at Tehachapi Mountain, with *C. carnana* from *Pseudotsuga macrocarpa* in the San Gabriel Mountains, and from both hosts in the San Bernardino Mountains The larvae are bright green, larger than other *Argyrotaenia* and not superficially distinguishable from those of *C. retiniana*.

Choristoneura rosaceana (Harris)

Lozotaenia rosaceana Harris, 1841, Rep. Insects Mass. Inj. Veg., p. 37.
Choristoneura rosaceana; Freeman, 1958, Canad. Entomol. 90, Suppl. 7:39.

The Oblique-banded Leaf-roller is widespread over much of the United States and southern Canada, and the larvae are polyphagous on a variety of trees and shrubs (Powell 1964). In the Forest Lepidoptera of Canada survey, more than 70 species of host plants were recorded, and among the 20 for which there were more than 10 collections, Salicaceae (42%) and Betulaceae (17%) predominate. However, here were 28 records for *Picea* and *Tsuga* among 1,000+ records (Prentice 1965).

In our survey, *C. rosaceana* was reared once, from *Pseudotsuga menziesii* in Napa Co. In California this insect is restricted to lower elevations, seldom occurring above 1,200 m, so it is sympatric with *C. carnana* through much of its range but rarely occurs in areas of *Abies* and *C. retiniana*.

In younger instars the larvae are similar to several other conifer-inhabiting tortricids, green with a black head and thoracic shield. Mature larvae are dark green, much larger than corresponding individuals of most other tortricid genera. The head and prothoracic shield are brown with black markings.

Archips argyrospila columbiana (McDunnough)

Retinia argyrospila Walker, 1863, List Lepid. Brit. Mus. 28:373.
Archips argyrospila; Fernald, 1903, in Dyar, U.S. Natl. Mus. Bull. 52:480.
Cacoecia columbiana McDunnough, 1923, Canad. Entomol. 55:167.
Archips argyrospila columbiana; McDunnough, 1939, Mem. South. Calif. Acad. Sci. 2:56.

Populations referred to *Archips argyrospila*, which was described originally from Georgia, have been shown on the basis of pheromone chemistry (Roelofs and Comeau 1969) and larval morphology (MacKay 1962) to represent a complex of species. McDunnough's name, *columbiana*, was based on a pale female specimen from British Columbia; whether California populations are referable to this name in a subspecific context is not certain, but comparable variants occur in California populations. MacKay (1962) described two larval forms in California, and it is possible that one represents an introduction into agricultural areas of the eastern Fruit-tree Leaf-roller. A detailed study emphasizing pheromone-based differences is needed to clarify understanding of relationships among western populations.

A. argyrospila in native habitats in California typically have a generally olivaceous hue in the adults, rather than a rust or reddish color as in populations of the eastern United States and Rocky Mountains; the female is much paler than the male, with the forewing markings washed out; and the last instar larva has the dorsum conspicuously darker than the rest of the body. The dark larval dorsum, which is due to dense, black spicules, is not developed in the eastern Fruit-tree Leaf-roller nor in some California populations in agricultural areas (MacKay 1962).

A. argyrospila everywhere is polyphagous; in California larvae are most often found on *Quercus* (Fagaceae) in native situations but are recorded on a diversity of Rosaceae and other plants, including *Pseudotsuga* (Powell 1964). During our surveys for *Choristoneura*, we reared *A. argyrospila* from *P. menziesii* in the Siskiyou-Trinity region and North Coast Range, and on both *P. macrocarpa* and *Abies concolor* in the San Bernardino Mountains

Volney *et al.* reared this species from *P. menziesii* and *A. concolor* at Boulder Creek, Siskiyou Co. (UCB specimens).

In younger instars the larvae are green, lacking the dark dorsum, with the head and thoracic shield black; thus they are not distinguishable superficially from several other tortricids.

Archepandemis coniferana Mutuura

Parapandemis borealis; (not Freeman, 1965) Furniss and Carolin, 1977, USDA Misc. Publ. 1339:174.
Archepandemis coniferana Mutuura, 1978, Canad. Entomol. 110:572.

This species was described from British Columbia, and it ranges south to northeastern California. Prior to 1978, populations were recognized but thought to be conspecific with *A. borealis* (Freeman), an eastern species. During the 1974 outbreak of *Choristoneura retiniana* in the Modoc region, larvae were collected from late May to early July on *Abies concolor* at 8 of 9 sampling sites, at very low incidence (n= 22 of 5,000+ moths reared) (Powell and Donahue 1974, reported as *Parapandemis borealis*). However, we did not detect *A. coniferana* in much smaller larval samples in that area in 1970 and 1980, nor elsewhere during our survey. The larvae are dark brown and not easily mistaken for any other species, at least in later instars. It seems likely that this is the species reported as an undescribed *Choristoneura* in Oregon by Carolin (1980).

Syndemis afflictana (Walker)

Sciaphila afflictana Walker, 1863, List. Lepid. Brit. Mus. 28:237.
Syndemis afflictana; Freeman, 1958, Canad. Entomol., 90, Suppl. 7: 43.

Although specimens from California of this species are rare in collections, because the adults fly in early spring, numerous males were attracted to synthetic pheromone bait deployed for *Choristoneura retiniana* early in the season at two sites in Plumas Co. and at Blodgett Forest, El Dorado Co., in 1982.

Walsingham (1879) reared *S. afflictana* from larvae taken from "fir" at Mt. Shasta, California, in August. Hostplant records from eastern North America (Powell 1964, Chapman and Lienk 1971) and for the closely related *S. musculana* (Hübner) in Europe (Bradley *et al.* 1973), indicate that *Syndemis* is a general feeder. However, 10 of 13 records in the Forest Lepidoptera of Canada survey were *Abies* (Prentice 1965), and the known localities in California, together with Walsingham's record, suggest that *Abies* is the primary hostplant in northern and western populations. There is an isolated population in coastal San Luis Obispo Co. that feeds on *Pinus radiata* (JAP 80C8).

According to Chapman and Lienk (1971), penultimate instar larvae are found in August and early September and full grown larvae in mid-September and October in New York; the mature larvae overwinter in shelters of dead leaves; and pupation occurs in early spring without further feeding. Prentice (1965) lists larval collections in May and June and again in August and September in Canada; perhaps at northern latitudes feeding is not completed in fall. If the life cycle in California is similar to that of New York populations, larvae of *Syndemis* would be seasonally delayed compared to those of *Choristoneura*, explaining why our larval survey failed to detect them.

The larvae are tan or yellowish with a dorsal line and broad, irregular, dorsolateral stripes of dark brown that tend to fade out posteriorly. The head and thoracic shield are amber, the latter with prominent black, lateral spots posteriorly (Chapman and Lienk 1971).

Clepsis persicana forbesi Obraztsov

Craesia persicana Fitch, 1856, Trans. N. Y. State Agric. Soc. 16:357.
Clepsis persicana; Freeman, 1958, Canad. Entomol. 90, Suppl. 7:56.
Clepsis forbesi Obraztsov, 1962, Amer. Mus. Novit. 2101:22.
Clepsis persicana forbesi; Powell, 1964, Univ. Calif. Publ. Entomol. 32:239.

In eastern North America this insect is known as the White-triangle Tortrix and is a general feeder that is sometimes an orchard defoliator. The western populations have the forewings more brightly colored, with the white costal triangle and other markings less well defined. Obraztsov considered *forbesi* to be a distinct species, but there is a shift to the western form at about the continental divide, with no evidence of sympatry of the two forms, and intermediates occur in that area. Records for the two races are combined in Forest Lepidoptera of Canada (Prentice 1965): more than 500 collections from 40 host species indicate that Salicaceae (43%) and Pinaceae (45%) are preferred. There are records for *Picea*, *Abies*, *Pseudotsuga*, *Larix*, and *Pinus*.

We detected larvae of *C. persicana* twice on *Abies concolor* in Siskiyou Co. These are the only known host records in California, but there are collections of adults from lower elevations than *Abies* occurs (Powell 1964 and unpubl. data). Stevens *et al.* (1984) list *Pseudotsuga* as a host tree of *C. persicana* in association with *Choristoneura occidentalis* but omit discussion of this species and its larval diagnosis from the text and keys.

The mature larva of *C. p. persicana* is yellow-green without markings except for a small dark dash anterolaterally on the head (Chapman and Lienk 1971).

Tortricidae: Sparganothini

Most members of this tribe are polyphagous, including the two species encountered in association with conifer-feeding *Choristoneura*.

Sparganothis senecionana (Walsingham)

Oenectra senecionana Walsingham, 1879, Illus. Lepid. Het. Brit. Mus. 4:17.
Sparganothis senecionana; Fernald, 1903, *in:* Dyar, U.S. Natl. Mus. Bull. 52:477.

Misidentification of the type specimen of *Sparganothis inconditana* (Walsingam), which is a synonym or race of the widespread *S. xanthoides* (Walker) (Powell unpubl. data), led to the erroneous placement of *S. senecionana* as a synonym of *inconditana* for nearly a century. *S. senecionana* was described originally from northern California; it occurs widely over the southwestern United States. Populations occur from seacoast beaches to above timberline in California and are markedly polytypic in color and forewing pattern.

There is a single annual generation; first instar larvae estivate and hibernate in silken hibernacula. In early spring they disperse and feed on all kinds of plants, particularly low growing, perennial herbs and shrubs. They also feed on conifers, sometimes in numbers, and we detected *S. senecionana* on *Abies magnifica* at Lake Tahoe at two sites and on both *A. concolor* and *Pseudotsuga menziesii* in the Salmon Mountains, Siskiyou Co. It also was reared by Volney *et al.* at Mt. Ashland from *A. concolor* (UCB specimens).

The larvae are dark green with the pinacula conspicuously paler and have a black head and thoracic shield. They are not easily distinguished from several other tortricines, including *Choristoneura retiniana*, in early instars.

Amorbia cuneana (Walsingham)

Hendecastema cuneanum Walsingham, 1879, Illus. Lepid. Het. Brit. Mus. 4:4.
Amorbia cuneana; Fernald, 1882, Trans. Am. Entomol. Soc. 10:19.

This species was described originally from Mt. Shasta, California; it is widely distributed in coastal and Transition Zone areas in California and Oregon. Pheromone studies indicate that there are two morphologically similar species in urban and agricultural situations (Bailey *et al.* 1988), one of which probably is *A. essigana* Busck. Populations in native montane habitats seem to be conspecific with Walsingham's species. A wide variety of native and introduced plants serve as hosts; in natural communities evergreen shrubs such as *Heteromeles* (Rosaceae) and *Arctostaphylos* (Ericaceae) seem to be preferred, providing smooth leaves that are selected for oviposition and overwintering sites in larval shelters between mature leaves.

Volney *et al.* reared *A. cuneana* at Mt. Ashland from *Abies concolor* (UCB specimen), and we found one larva on *Pseudotsuga menziesii* in Mendocino Co., with *Choristoneura carnana*. Larvae of younger instars resemble those of other tortricines and differ markedly from those of *C. carnana*. Mature larvae are pale olive green with a tan head and thoracic shield having characteristic dark, lateral markings. *A. cuneana* is primarily a lowland species that rarely ranges to elevations where it is sympatric with *C. retiniana*.

Pyralidae

Dioryctria pseudotsugella Munroe

Dioryctria pseudotsugella Munroe, 1959, Canad. Entomol. 91:68.

This species was described from British Columbia, where most of the type series was reared from Douglas-fir, but two specimens from *Abies* were included.

Mutuura and Munroe (1973) found that the species long known in the literature as the Spruce Coneworm had been misidentified as *D. reniculella* (Grote). They recognized the latter name as a subjective synonym of *D. abietivorella* (Grote), and they proposed the name *D. reniculelloides* for the Spruce Coneworm. *D. pseudotsugella* and *D. reniculelloides* are closely related and are reported to feed primarily on Douglas-fir and *Picea*, respectively, although both are recorded on various Abietoideae (Mutuura and Munroe 1973). While sympatric populations associated with these hosts in Canada may be distinguishable, we cannot differentiate two species in material reared from *Pseudotsuga* and *Abies* in California. The species illustrated as *D. pseudotsugella* by Stevens *et al.* (1984: fig. 29) is the prevalent form in California, although there is variation in series of reared specimens. We consider all these southern populations to represent *D. pseudotsugella*.

The larvae feed primarily in foliage, but evidently also in flowers and green cones, and sometimes they are numerous. During our sampling, *Dioryctria* varied in abundance in accord with *Choristoneura*; when the latter was at low density we never found *D. pseudotsugella* to be numerous. During monitoring of the 1974 outbreak of *C. retiniana* on *Abies concolor* in the Modoc area, we reared more than 2,000 *D. pseudotsugella*. This was the most frequently reared non-*Choristoneura* during our survey. It was collected from *A. concolor* in Modoc, the Trinity Mountains, and Tehachapi Range; from *Pseudotsuga menziesii* in the Mt. Shasta region and Coast Range in Napa and Lake Cos.; and from *P. macrocarpa* in the San Gabriel Mountains.

The larvae are predominantly brick red, similar in color to those of *Choristoneura carnana*. Larvae of *Dioryctria* are pale dorsally with distinctive longitudinal bands of gray dorsolaterally, usually bordered by irregular whitish lines, and there are conspicuous, sclerotized dark rings around the SD1 setae on the mesothorax and 8th abdominal segments. We did not observe the pale form illustrated by Stevens *et al.* (1984: fig. 27). According to Mutuura and Munroe (1973), preserved larvae of *D. pseudotsugella* from British Columbia and Montana appear to lack the longitudinal stripes, but this integumental color was not lost in our preserved samples.

Geometridae

Neoalcis californiaria (Packard)

Boarmia californiaria Packard, 1869 (1871?), Proc. Boston Soc. Nat. Hist. 13:388.
Neoalcis californiaria; McDunnough, 1920, Bull. Canad. Dept. Agric. Entomol. 18:21.

This is a commonly collected, solitary looper that feeds on various conifers along the Pacific Coast. In British Columbia, *Pseudotsuga* with 38% of more than 550 collection records, and *Tsuga* with 31%, were the most frequently detected hosts. Various other conifers, including *Thuja* (Cupressaceae) with 15% of the records, and *Pinus*, were used (Prentice 1963).

In California, the species occurs in north coastal areas and inland at moderate elevations, up to about 1,200 m, below the occurrence of *Abies concolor*. *N. californiaria* has been reared from Coast Redwood, *Sequoia sempervirens* (Taxodiaceae), in Santa Cruz Co. (JAP 74D28), from *Cupressus sargenti* (Cupressaceae) in Marin Co. (JAP 92D26), and we found the larvae in numbers on Douglas-fir at two sites in Napa Co. *N. californiaria* sometimes feeds on non-coniferous shrubs; we collected larvae (n=4 reared) on *Ceanothus thrysiflorus* (Rhamnaceae) growing with *Sequoia* in Marin Co. (JAP 78D29) and Monterey Co. (JAP 89C78) and on *C. sorediatus*, with *Pinus sabiniana* the only conifer nearby, in Yolo Co. (JAP 92C33). The larvae are slender, brown with irregular dark markings, and when disturbed they hold themselves rigid, erect from the substrate and appear twig-like. The adults fly primarily in September and October; specimens reared from *Pseudotsuga* emerged from July to early October.

Cochisea sonomensis McDunnough

Cochisea sonomensis McDunnough, 1941, Canad. Entomol. 73:75.

This species is endemic to the North Coast Ranges of California, from Trinity to Santa Clara Cos. (Rindge 1975). Rindge reported one rearing record from *Pinus*, and pinyon pine is the host of the closely related species, *C. recisa* Rindge (Comstock *et al.* 1958, reported as *C. sonomensis*, Rindge 1975). We found larvae of *C. sonomensis* on *Pseudotsuga* in Napa and Siskiyou Cos., in May and June, in association with *Choristoneura carnana*.

The larvae are dark green, bulky loopers with longitudinal, dorsolateral stripes and diagonal lines which render a diamondback pattern that is reminiscent of *Cupressus*-feeding caterpillars. There is a rare gray form of the larva in *C. recisa* (Comstock *et al.* 1958). The adults emerge in the fall.

Gabriola dyari Taylor

Gabriola dyari Taylor, 1904, Canad. Entomol. 36:256.

The typical race of *G. dyari* occurs along the Pacific Coast from southern Alaska to central California (Rindge 1974). Various conifers are recorded as hostplants (Sugden 1968); in British Columbia, *Tsuga heterophylla* with 47% and *Pseudotsuga* with 32% of more than 450 collections were most frequent among 14 hosts reported (Prentice 1963). The lists includes *Pinus* and *Thuja* (Cupressaceae), each with about 4.5% of the records.

We reared *G. dyari* on *Pseudotsuga menziesii* in Lake and Napa Cos. Flight records at Berkeley indicate that the species is multivoltine. The central coast subspecies, *G. dyari pruina* Rindge, has been collected in every month except July (Rindge 1974). It feeds on *Pinus radiata* at the type locality, Cambria, San Luis Obispo Co. (JAP 80C9).

The larva of *Gabriola dyari* is reddish brown with pale segmental markings dorsally and has a slightly bilobed head capsule and prominent bilobed, lateral pinacula on the abdominal segments. According to Sugden (1968), larvae in British Columbia have two color forms: one is buff flecked with brown and black, dorsally with an indistinct pale line and "+"-shaped marks and black, diagonal, subdorsal lines; the other is pale buff with orange markings rather than dark. When disturbed, our larvae assumed a curled posture and resembled a bird dropping, or when young, a staminate cone of Douglas-fir.

Nepytia umbrosaria nigrovenaria (Packard)

Cleora umbrosaria Packard, 1873 (1874?), Proc. Boston Soc. Nat. Hist. 16:23.
Cleora nigrovenaria Packard, 1876, U.S. Geol. Surv. Terr. 10:454.
Nepytia nigrovenaria; Hulst, 1896, Trans. Amer. Entomol. Soc. 23:344.
Nepytia umbrosaria nigrovenaria; Barnes and McDunnough, 1917, Check List Lepid. Boreal Amer., p. 115.

Larvae of *N. nigrovenaria* occur as scattered individuals on Douglas-fir and other conifers, including *Pinus* and *Thuja* (Cupressaceae), in the Pacific states from British Columbia to California (Furniss and Carolin 1977). Older reports such as Prentice (1963) may include records referable to *N. freemani* Munroe, which was described in 1963.

This species is widespread in California; adults often came to lights at our *Choristoneura* sampling sites, including higher elevation localities, but we found larvae only on *Pseudotsuga menziesii* at two coastal localities in Marin and Sonoma Cos., and on *Pinus radiata* at Cambria, San Luis Obispo Co. (JAP 80C9). The larvae are elongate, slender, reddish brown with cream colored, longitudinal, dorsolateral stripes and venter.

Enypia packardata Taylor

Enypia packardata Taylor, 1906, Canad. Entomol. 38:403.

This species is a Pacific Coast counterpart of the closely related *E. griseata* Grossbeck, which ranges through the Rocky Mountains and eastern portions of Oregon, Washington, and British Columbia (Evans 1960). The hostplants of *E. packardata* include various Abietoideae and rarely *Thuja*, *Chamaecyparis* (Cupressaceae), and *Pinus* (Evans 1960, Prentice 1963). In British Columbia, *Tsuga* with 47%, and *Pseudotsuga* with 37%, were the prevalent hosts among nearly 600 collections in the Forest Lepidoptera Survey.

We reared *E. packardata* from Douglas-fir at Angwin, Napa Co., and from *Abies concolor* at Tehachapi Mtn. The species occurs throughout the ranges of these two trees in California. The larva has an orange-brown head capsule, and the body is green with a narrow, pale dorsal line and broad, whitish subdorsal stripes that are edged with dark green (Evans 1960).

Saturniidae

Hyalophora euryalus (Boisduval)

Saturnia euryalus Boisduval, 1855, Bull. Soc. Entomol. Fr. 3 (ser. 2):132.
Hyalophora euryalus; Bovier, 1936, Mem. Mus. Nat. Hist. Natl. (n.s.) Paris 3:270 (as *rubra*).

This large red silkmoth is widespread in the Pacific states and ranges from sea level to 2,700 m elevation in California. The larvae are polyphagous on non-coniferous shrubs and trees (Collins 1973, Ferguson 1972) but often seem to specialize within a given population, for example on *Rhamnus* and *Ceanothus* (Rhamnaceae) in the Coast Range chaparral and *Salix* in the Sacramento River delta.

Larvae have been collected on *Pseudotsuga* in the Klamath-Cascade region of California by T. Koerber (Collins, 1973), and a photograph of the caterpillar eating Douglas-fir is shown by Furniss and Carolin (1977). We found young *H. euryalus* larvae and reared them to maturity on old foliage of *Pseudotsuga menziesii* on two occasions in Napa Co. and at Blodgett Forest, El Dorado Co. In contrast to all other species that we reared, *H. euryalus* fed exclusively on the previous seasons' needles and avoided new growth, even after consuming available older foliage. The larvae are green with paired dorsal tubercles, those of the thoracic segments larger, yellow, and banded with black.

Arctiidae

Lophocampa argentata (Packard)

Halesidota argentata Packard, 1864, Proc. Entomol. Soc. Phila. 3:129.
Lophocampa argentata; Franclemont, 1983, in: Hodges *et al.* Check List Lepid. Am. N. of Mexico, p. 118.

The Silver-spotted Tiger Moth occurs in the Pacific states from British Columbia to California. It feeds primarily on *Pseudotsuga* but also on other conifers including *Pinus* and *Thuja* (Cupressaceae), according to Furniss and Carolin (1977), who state that the species is a recurrent defoliator in coastal areas.

We encountered the larvae only in low numbers on Douglas-fir in the North Coast Range. In this region, development is completed in May and June, and adults emerge in July. Furniss and Carolin report that the immature larvae overwinter gregariously in loose, tentlike webs.

Noctuidae

Syngrapha celsa (H. Edwards)

Plusia celsa H. Edwards, 1891, Papilio 1:101.
Syngrapha celsa; Hampson, 1913, Cat. Lepid. Phalaenae Brit. Mus. 13:422.

This noctuid occurs from Alaska to California and east to Utah; in California it ranges along the North Coast and in montane areas the length of the state (Eichlin 1975), roughly corresponding to the distribution of *Abies concolor*. Stevens *et al.* (1984) state that *S. celsa* is well known as an occasional associate of *Choristoneura occidentalis* in the Pacific Northwest, but they do not specify hostplants. In British Columbia the species was rare during the Forest Lepidoptera Survey, feeding on *Abies, Picea, Tsuga*, and *Pinus* (Prentice 1962).

Syngrapha celsa was reared from larvae taken during the 1974 Modoc budworm outbreak on *Abies concolor* at two localities (Powell and Donahue 1974). We also collected but did not successfully rear larvae thought to be this species, on *A. concolor* at McBride Springs, Mt. Shasta.

The larvae are semiloopers, lacking the abdominal prolegs of segments 3 and 4, and are bright green with dorsolateral white stripes. The integument is thickly covered with dark microspines, which form middorsal and subdorsal, longitudinal bands (Eichlin and Cunningham 1978).

Achytonix epipaschia (Grote)

Homohadena epipaschia Grote, 1882, Trans. Kansas Acad. Sci. 8:49.
Achytonix epipaschia; McDunnough, 1937, Canad. Entomol. 69:61.

This species evidently is a minor element in the conifer-feeding community in western North America, but in conjunction with other noctuids can cause defoliation of Douglas-fir (Furniss and Carolin 1977). A related species, *A. praeacuta* (Smith), was considered to be synonymous by Furniss and Carolin (and presumably by Stevens *et al.* 1984), although it is regarded as a distinct species by taxonomists (Crumb 1956, Prentice 1962, Franclemont and Todd 1983). *A. epipaschia* (including *praeacuta*) feeds on conifers, primarily *Pseudotsuga* according to Furniss and Carolin, but in British Columbia and Alberta 95% of 178 collections of *A. praeacuta* were from *Abies* (Prentice 1962). We reared *Achytonix* only once, from Douglas-fir in the Trinity Mountains, where it fed with *Choristoneura carnana.*

The larva is bright bluish green with dorsal and dorsolateral white lines, a dark line at the spiracles, and a subventral yellowish stripe; the head is green (Crumb 1956).

Feralia deceptiva McDunnough

Feralia deceptiva McDunnough, 1920, Canad. Entomol. 52:162.

This striking green noctuid was described from British Columbia and occurs southward to the North Coast Range and Sierra Nevada in California. The distribution is not well documented because the adults fly in early spring and are rarely collected. The larvae feed on *Pseudotsuga menziesii* in Washington (Crumb 1956), and we collected one on Douglas-fir at Blodgett Forest, El Dorado Co.

The larva is green, including the head, with middorsal, dorsolateral, and ventrolateral white stripes. The 8th abdominal segment is slightly broadened and round-humped dorsally (Crumb 1956).

We also reared *Feralia februalis* Grote from *Pseudotsuga*, at Alpine Lake, Marin Co. (JAP 78D25), where *Choristoneura* has not been recorded. This species is a general feeder on hardwoods (Crumb 1956, Prentice 1962, UCB unpubl. data), so the larva on Douglas-fir may have been incidental.

Egira crucialis (Harvey)

Xylomiges crucialis Harvey, 1875, Bull. Buffalo Soc. Nat. Sci. 2:277.
Egira crucialis; Franclemont and Todd, 1983, *in* Hodges (ed.), Check List Lepid. Am. N. of Mexico, p. 151.

Egira crucialis is widespread in the Pacific states, and the larvae are recorded from various hardwoods (Crumb 1956, Prentice 1962). However, we found them on *Pseudotsuga menziesii* at Angwin, Napa Co. The association did not appear to be incidental; young larvae (n=ca 12) tied needles of the new season growth into tight shelters, and feeding continued to maturity. The pupae overwinter and appear to be capable of carrying over in diapause to a second or later season (Powell 1987).

In early instars the larva of *E. crucialis* is olive green, paler ventrally, with thin, whitish dorsal and subdorsal lines; the head, thoracic shield and pinacula are shining black. Full grown larvae are dull olive tinged with reddish, and have distinct white dorsal and dorsolateral lines edged darker and a reddish or chestnut colored head.

Egira simplex (Walker)

Xylomiges simplex Walker, 1865, Cat. Lepid. Brit Mus. 32:618.
Egira simplex; Franclemont and Todd, 1983, *in* Hodges (ed.) Check List Lepid. Am. N. of Mexico, p. 151.

Egira simplex is widespread in the western United States and is similar in appearance to *E. crucialis*, which was considered to be a synonym in early literature (Dyar 1903). *E. simplex* was recorded feeding on *Pseudotsuga* from two collections in British Columbia

(Prentice 1962), and we found larvae on *P. menziesii* near Angwin, Napa Co. (n=2+, 1 adult reared). The species was reported as a general feeder on hardwoods as well as *Pseudotsuga* by Furniss and Carolin (1977) and Stevens *et al.* (1984). However, these authors did not mention *E. crucialis*, possibly because they considered the two *Egira* to be synonymous.

Early instar larvae of *E. simplex*, which are needle-tiers, are olive green with a black head capsule. Late instars are green, matching Douglas-fir foliage, with whitish, more or less complete, dorsal, dorsolateral, and subventral lines, blackish triangular marks adjacent to the DL line, and black spiracles in the SV stripe. In the penultimate instar, the head is brown with dark lines above the epicranial sutures directed to the posterior margin, and in the last instar the head is unmarked reddish.

Anomogyna mustelina (Smith)

Hadena mustelina J. B. Smith, 1900, Proc. U. S. Natl. Mus. 22:469.
Anomogyna mustelina; McDunnough, 1938, Mem. So. Calif. Acad. Sci. 1:66.

This gray noctuine, which was originally described from California, is widespread in the Pacific Northwest. Crumb (1956) reported the hosts to be mainly *Abies* but also *Pseudotsuga* and pine in Oregon and Washington, while in British Columbia Prentice (1962) listed 226 larval collections, of which 46% were from *Pseudotsuga*, 28% from *Tsuga*, and only 10% from *Abies*. The list includes 9 records from *Thuja* (Cupressaceae) and one from *Pinus*.

Anomogyna mustelina was collected once during preliminary survey of the 1974 Modoc budworm outbreak on *Abies concolor*. There seem not to be any previous records on White Fir. *A. mustelina* evidently also feeds on *Pseudotsuga* in California, as it occurs in the Coast Range (Lake Co.) at lower elevation than does *A. concolor*.

The mature larva is dull purplish gray with a black-edged, whitish dorsal line and a whitish subdorsal band subtended by a broad purplish stripe and a broad, yellow subventral stripe. The head is whitish with brown or reddish markings (Crumb 1956). Stevens *et al.* (1984) state that the larva overwinters and feeds in the opening buds in spring, but in British Columbia *A. mustelina* overwinters as a pupa, according to Prentice (1962).

CONCLUSIONS

Two members of the spruce budworms complex feed on Abietoideae in California: *Choristoneura retiniana* on *Abies*, and *C. carnana* primarily on *Pseudotsuga*. At least 45 other species of Lepidoptera, members of 10 families, share this larval habitat, making up a caterpillar community that varies in relative numbers of species and individuals from place to place and from season to season. About 70% of the associated species are strictly conifer-feeders, of which most are Abietoideae specialists and the rest conifer generalists that use *Abies* or *Pseudotsuga* occasionally to frequently. The remaining 30% consists of polyphagous Lepidoptera that usually are not numerous on conifers but on a local basis in low density populations may make up an appreciable proportion of the larvae.

In high density populations of *Choristoneura*, associated species may be low, 10% or fewer of the defoliators, but surveys in Oregon by Carolin (1980), in Idaho by Markin (1982), and our observations indicate that often there are parallel trends of increase. As numbers of budworm rise, so do those of other Lepidoptera, and their populations sometimes reach 25-50% in the guild. The representation of *Choristoneura* varies dramatically, from 0 to 75% or more, at low population levels (< 10 larvae/search hr), which is the situation in most of California. In these circumstances, detection and monitoring the development of the budworm populations can be extremely difficult.

Many of the external feeding caterpillars, such as Arctiidae, Noctuidae, and Geometridae, are easily distinguished from *Choristoneura*, but other larvae that live within shelters often are quite similar in appearance, particularly in early instars. In general, many caterpillars on *Pseudotsuga* are reddish brown with longitudinal white markings, and although resembling *C. carnana* in color, most are easily differentiated by structural characters. In contrast, many species that inhabit *Abies* are green, similar to *C. retiniana* by cryptic resemblance to the fir foliage. Included are several tortricids that are structurally quite similar as well. The mining larvae of *Epinotia meritana, E. trossulana,* and *Coleotechnites* species can be mistaken for early instar *Choristoneura*. We doubt that anyone can distinguish with confidence the early instar larvae of several tortricine species from those of *C. retiniana* in the field (e.g., *Acleris gloverana, Archips argyrospila, Clepsis persicana, Sparganothis senecionana*). Moreover, late instars of some *Argyrotaenia* species are superficially indistinguishable from smaller individuals of last instar *C. retiniana.*

Monitoring any population of *Choristoneura*, but particularly *C. retiniana*, should include laboratory examination with good optics of well preserved exemplars of each larval morph. We never found fewer than six species associated with *Choristoneura* in new season foliage of *Abies* or *Pseudotsuga* in California. They made up 50-100% of the caterpillar community at most sites. We believe that the significance of associated species has been underestimated by many budworm researchers.

LITERATURE CITED

Bailey, J.B., K.N. Olsen, L.M. McDonough, and M.P. Hoffmann. 1988. Possible new race of *Amorbia cuneana* discovered in avocado. Calif. Agric. 42(5):11-12.

Bradley, J.D., W.G. Tremewan, and A. Smith. 1973. British Tortricoid Moths. Cochylidae and Tortricidae: Tortricinae. The Ray Society, London; 251 pp.

Braun, A.F. 1921. Notes on Microlepidoptera, with descriptions of new species. Entomol. News 32:8-18.

Burdick, D.J., and J.A. Powell. 1960. Studies on the early stages of two California moths which feed in the staminate cones of digger pine (Lepidoptera: Gelechiidae). Canad. Entomol. 92:310-320.

Carolin, V.M. 1980. Larval densities and trends of insect species associated with spruce budworms in buds and shoots in Oregon and Washington. USDA For. Serv. Res. Pap. PNW-273; 18 pp.

Chapman, P.J., and S.E. Lienk. 1971. Tortricid fauna of apple in New York (Lepidoptera: Tortricidae). N.Y. Agric. Exp. Stn., Geneva; 122 pp.

Clarke, J.F.G. 1934. Notes and new species of Microlepidoptera from Washington State. Canad. Entomol. 66:171-181.

Collins, M.M. 1973. Notes on the taxonomic status of *Hyalophora columbia* (Saturniidae). J. Lepid. Soc. 27:225-235.

Comstock, J.A., C. Henne, and F. Sala. 1958. The habits and life histories of *Cochisea sinuaria* B. and McD. and *Cochisea sonomensis* B. and McD. (Lepidoptera: Geometridae). Bull. So. Calif. Acad. Sci. 56:169-177.

Crumb, S.E. 1956. The larvae of the Phalaenidae. USDA Tech. Bull. 1135; 356 pp.

De Benedictis, J.A., D.L. Wagner, and J.B. Whitfield. 1990. Larval hosts of Microlepidoptera of the San Bruno Mountains, California. Atala 16:14-35.

Dyar, H.G. 1903. A list of North American Lepidoptera. U.S. Natl. Mus. Bull. 52; 722 pp.

Eichlin, T.D. 1975. A guide to the adult and larval Plusiinae of California (Lepidoptera: Noctuidae). Occas. Pap. Entomol. (Sacramento) 21:1-73.

Eichlin, T.D. and H.B. Cunningham. 1978. The Plusiinae (Lepidoptera: Noctuidae) of America north of Mexico, emphasizing genitalic and larval morphology. USDA Tech. Bull. 1567; vi + 122 pp.

Evans, D. 1960. A revision of the genus *Enypia* (Lepidoptera: Geometridae). Ann. Entomol. Soc. Amer. 53:560-574.

Ferguson, D.C. 1972. Bombycoidea: Saturniidae (Part). Moths Amer. N of Mexico; fasc. 20.2B, pp. 155-275; E. W. Classey and Wedge Entomol. Res. Found., London.

Franclemont, J.G., and E.L. Todd. 1983. Noctuidae. *In*: R. W. Hodges *et al.* (eds.): Check list of the Lepidoptera of America north of Mexico including Greenland, pp. 120-159; E.W. Classey, Ltd. and the Wedge Entomol. Res. Found., London.

Furniss, R.L., and V.M. Carolin. 1977. Western Forest Insects. USDA For. Serv. Misc. Publ. 1339; vii + 654 pp.

Keifer, H.H. 1933. California Microlepidoptera, VI. Calif. Dept. Agric. Mo. Bull. 22:351-365.

Lyon, R.L., C.E. Richmond, J.L. Robertson, and B.A. Lucas. 1972. Rearing diapause and diapause-free western spruce budworm (*Choristoneura occidentalis*) (Lepidoptera: Tortricidae) on artificial diet. Canad. Entomol. 104:417-426.

MacKay, M.R. 1962. Larvae of the North American Tortricinae (Lepidoptera: Tortricidae). Canad. Entomol. Suppl. 28; 182 pp.

Markin, G.P. 1982. Abundance and life cycles of Lepidoptera associated with an outbreak of the Western Spruce Budworm, *Choristoneura occidentalis* (Lepidoptera: Tortricidae) in southeastern Idaho. J. Kansas Entomol. Soc. 55:367-372.

Mutuura, A., and E. Munroe. 1973. American species of *Dioryctria* (Lepidoptera: Pyralidae). IV. The *schuetzeella* group and the taxonomic status of the spruce cone moth. Canad. Entomol. 105:653-658.

Obraztsov, N.S. 1964. *Griselda radicana* Heinrich, 1923 (Insecta: Lepidoptera): Proposed validation under the plenary powers. Z.N.(S). 1612. Bull. Zool. Nomencl. 21:144-145.

Obraztsov, N.S. 1965. On *Paedisca radicana* and the type species of the genus *Griselda* (Lepidoptera, Tortricidae, Olethreutinae). Amer. Mus. Novit. 2213; 16 pp.

Powell, J.A. 1962. Taxonomic studies on the *Acleris gloverana-variana* complex, the black-headed budworms (Lepidoptera: Tortricidae). Canad. Entomol. 94:833-840.

Powell, J.A. 1964. Biological and taxonomic studies on tortricine moths, with reference to the species in California (Lepidoptera: Tortricidae). Univ. Calif. Publ. Entomol. 32:1-317.

Powell, J.A. 1987. Records of prolonged diapause in Lepidoptera. J. Res. Lepid. 25:83-109.

Powell, J.A. and J.A. De Benedictis. 1995. Biological relationships: host tree preferences and isolation by pheromones among allopatric and sympatric populations of western *Choristoneura*. Part 2 this volume.

Powell, J.A., and V. Donahue. 1974. Monitoring the effects of Dylox on microlepidopterous larvae. (Unpublished report to USDA For. Serv.; 17 pp.)

Prentice, R.M. (compiler). 1958. Forest Lepidoptera of Canada recorded by the Forest Insect Survey. Vol. 1. Papilionidae to Arctiidae. Canad. Dept. Agric. Publ. 1034; 76 pp.

Prentice, R.M. (compiler). 1962. Forest Lepidoptera of Canada recorded by the Forest Insect Survey. Vol. 2. Nycteolidae, Noctuidae, Notodontidae, Liparidae. Forest Ent. and Path. Branch, Canad. Dept. For. Bull. 128; pp. 77-281.

Prentice, R.M. (compiler). 1963. Forest Lepidoptera of Canada recorded by the Forest Insect Survey. Vol. 3. Lasiocampidae, Thyatiridae, Drepanidae, Geometridae. For. Ent. and Path. Branch, Canad. Dept. For. Publ. 1013; pp. 283-543.

Prentice, R.M. (compiler). 1965. Forest Lepidoptera of Canada recorded by the Forest Insect Survey. Vol. 4. Microlepidoptera. Canad. Dept. For. Publ. 1142; pp. 545-840.

Rindge, F.H. 1974. A revision of the moth genus *Gabriola* (Lepidoptera, Geometridae). Am. Mus. Novit. 2550:1-24.

Rindge, F.H. 1975. A revision of the New World Bistonini (Lepidoptera, Geometridae). Bull. Am. Mus. Nat. Hist. 156:71-155.

Roelofs, W.L. and A. Comeau. 1969. Sex pheromone specificity: taxonomic and evolutionary aspects in Lepidoptera. Science 165:398-400.

Stevens, R.E. 1959. Biology and control of the pine needle-sheath miner, *Zelleria haimbachi* Busck (Lepidoptera, Hyponomeutidae). USDA Pac. SW For. and Range Exp. Stn. Tech. Pap. 30; 20 pp.

Stevens, R.E., V.M. Carolin, and G.P. Markin. 1984. Lepidoptera associated with western spruce budworm. USDA For. Serv., Agric. Handbook no. 622; 63 pp.

Sugden, B.A. 1968. Annotated list of forest insects of British Columbia. Part 14. Ennominae (Geometridae). J. Entomol. Soc. Br. Columbia 65:24-33.

Walsingham, Lord Th. de Grey. 1879. Illustrations of typical specimens of Lepidoptera Heterocera in the collection of the British Museum. Part IV. North American Tortricidae. British Museum, London; 84 pp.

Table 6.1. Foliage-feeding Lepidoptera reared from larvae collected on *Abies* and *Pseudotsuga* in California and southern Oregon during 1970-1982. (Numbers indicate total collections; localities in parentheses. Host preferences: A = Abietoideae, C = conifer, G = general feeder, Q = Quercus)

	Abies concolor	*Abies magnifica*	*Pseudotsuga menziesii*	*Pseudotsuga macrocarpa*	Host preference
GELECHIIDAE					
Coleotechnites sp.1	6 (5)				A
Coleotechnites sp.2	23 (17)	3 (2)	8 (8)		A
Coleotechnites sp.3				5 (4)	A
Telphusa sedulitella			2 (2)		Q
Chionodes abella	14 (10)		1 (1)		A
Chionodes periculella			4 (3)		A
Chionodes sabinianae	1 (1)				C
PLUTELLIDAE					
Yposolopha cervella			2 (2)		Q
YPONOMEUTIDAE					
Zelleria haimbachi	1 (1)		1 (1)	1 (1)	C
TORTRICIDAE					
Epinotia radicana	18 (12)	2 (2)	11 (7)		A
Epinotia meritana	4 (3)				A
Epinotia trossulana	4 (3)	1 (1)			A
Acleris gloverana	9 (7)	1 (1)	3 (3)		A
Argyrotaenia provana	8 (7)	1 (1)	2 (2)		A
Argyrotaenia citrana			1 (1)		G
Argyrotaenia dorsalana	17 (9)	1 (1)			A
Argyrotaenia lautana	2 (2)	1 (1)			A
Chor. rosaceana			1 (1)		G
Choristoneura carnana	5 (5)		23 (13)	4 (3)	A
Chor. retiniana	51 (27)	1 (1)			A
Chor. l. subretiniana	4 (3)	1 (1)			C
Archips a. columbiana	1 (1)		9 (8)	1 (1)	G
Archepan. coniferana	12 (8)				A
Clepsis p. forbesi	2 (2)				G
Spargano. senecionana	1 (1)	2 (2)	1 (1)		G
Amorbia cuneana			1 (1)		G
PYRALIDAE					
Dioryc. pseudotsugella	40 (14)		5 (5)	1 (1)	A
GEOMETRIDAE					
Neoalcis pseudotsugella			2 (2)		C
Cochisea sonomensis			2 (2)		C
Gabriola dyari			2 (2)		C
Neptyia umbrosaria			2 (2)		C
Enypia packardata	1 (1)		1 (1)		C
SATURNIIDAE					
Hyalophora euryalus			3 (2)		G

	Abies concolor	*Abies magnifica*	*Pseudotsuga menziesii*	*Pseudotsuga macrocarpa*	Host preference
ARCTIIDAE					
Lophocampa argentata			4 (3)		C
NOCTUIDAE					
Syngrapha celsa	3 (2)				A
Achytonix epipaschia			1 (1)		A
Feralia deceptica			1 (1)		A
Egira crucialis			2 (1)		G
Egira simplex			1 (2)		G?
Anomogyna mustelina	1 (1)				C
TOTAL RECORDS					
Associates	163	12	73	11	258
Choristoneura	60	2	23	4	89

Table 6.2. Percentage representation of *Choristoneura* and Associated Lepidoptera in Larval Collections, 1979-1982, Pooled by Region

Area	n Sites	n Individual	% *Choristoneura retiniana*	% *Choristoneura carnana*	% Other Lepidoptera (spp)	*Choristoneura* per search hour
Mt. Shasta - Modoc 1980	5	76	23.7	-	76.3 (6)	1.5
Lake Tahoe 1979-80	3	33	30.3	-	69.7 (8)	< 1
Blodgett Forest 1979-81	3	72	15.3	4.2	75.9 (7)	< 1
Salmon Mts., Siskiyou 1980	3	61	1.6	52.5	45.9 (12)	2-5.5
Trinity Mts. 1980	2	46	-	80.4	19.6 (7)	7.5
Lake-Napa Cos. 1979-80	4	184	-	48.4	51.6 (14+)	1-4.5
Tehachaipi Mt. 1980-82	1	455	87	-	13 (8)	40-50
San Gabriel Mts. 1980-81	3	47	14.9	21.3	63.8 (6)	1.5

Appendix 6.1. Lepidoptera Associated with Conifer-Feeding *Choristoneura* Species in California and Oregon. All collections California except as indicated Oregon. (See methods for explanation of lot numbering system.)

GELECHIIDAE:

Coleotechnites sp. 1	Abies concolor
Sugar Hill, Modoc Co.	70F35
Buck Creek Ranger Station, Modoc. Co.	70F38
Mill Crk., Warner Mts., Modoc. Co.	74G35.3
Soup Crk., Warner Mts., Modoc Co.	74G41.3, G59.2
Benton Mdws., Warner Mts., Modoc Co.	74G51.2
S. Deep Crk., Warner Mts., Modoc Co.	74G55.3
Blodgett Forest, El Dorado Co.	79F28, F39, 80F112, 81F117
Tehachapi Mt. Park , Kern Co.	82F15.2
Coleotechnites sp. 2 Abies concolor	
Fandango Pass, Warner Mts., Modoc Co.	70F68
2 mi. NW Fort Bidwell, Modoc Co.	70F83.1
9 mi. N Adin, Modoc Co.	74F25.1
9 air km NE Adin, Modoc. Co.	80F99
8 air km E Adin, Modoc Co.	80F105.1
4 km SW Etna, Siskiyou Co.	80F74
5 air km SW Etna, Siskiyou Co.	80F65
McBride Spr., Mt. Shasta, Siskiyou Co.	74F33, F54
5 mi. W McCloud, Siskiyou Co.	74F28
Snowmans Spr., Mt. Shasta, Siskiyou Co.	80F88
Snowmans Summit, Mt. Shasta, Siskiyou Co.	80F91
Mt. Pinos, Mc Gill, Cpgd., Ventura Co	81F115
Mt. Pinos, Mt. Pinos Cpgd., Ventura Co.	81F113
Blue Ridge Summit, Los Angeles Co.	81F109

Coleotechnites sp. 2	Abies magnifica
Ashland Ski Bowl, Jackson Co., Ore.	70G8
nr. Donner Lake, Nevada Co.	79G4
Ward Creek, 2. mi. W Hwy. 89, Placer Co.	79G6
2 mi. S Tahoe City, Ward Crk., Placer Co.	79G8
Kaspian Rec. Area, Placer Co.	79G12, 80G14
Coleotechnites sp. 2	Pseudotsuga menziesii
Juanita Lake, Siskiyou Co.	80F80
11 air km SW MacDoel, Siskiyou Co.	80F82.1
Lake Almanor, Plumas Co.	70G18
Johnsonville, Eureka Crk., Plumas Co.	70G20
3 mi. N Branscomb, Mendocino Co.	80E53
1.5 mi. W Branscomb, Mendocino Co.	80E56
Blodgett Forest, El Dorado Co.	79F37
Coleotechnites sp. 3	Pseudotsuga macrocarpa
4 mi. S. Crystal Lake, Los Angeles Co	80F40
Manker Flat Cpgd., San Bernardino Co.	80F31
3 air km N Mt. Baldy P. O., San Bernardino Co.	80F35
Mt. San Antonio, Manker Flat, San Bernardino Co.	81F107.2
Angelus Oaks, San Bernardino Co.	82F20
Telphusa sedulitella (Bsk.)	Pseudotsuga menziesii
Alpine Lake, Marin Co.	78D26
Angwin, Napa Co.	79E55
Chionodes abella (Bsk.)	Abies concolor
25 air mi. E Medford, Jackson Co., Ore.	70G12
Fandango Pass, Warner Mts., Modoc Co.	70F14
2 mi. N Fort Bidwell, Modoc Co.	70F83.2
Fandango Pk., Warner Mts., Modoc Co.	70G13.4
Soup Cr., Warner Mts., Modoc Co.	74G23.3
Snell Spr., Warner Mts., Modoc Co.	74G29.2

10 air km SE Trinity Ctr., Trinity Co.	8OF48.1
Scott Mt., Trinity Co.	80F55.1
McBride Spr., Mt. Shasta, Siskyou Co.	74F32
Snowmans Summit, Mt. Shasta, Siskiyou Co.	80F90
Blodgett Forest, El Dorado Co.	80F113
Tehachapi Mt. Park, Kern Co.	80F26.2, 81F102
Chionodes abella (Bsk.)	Pseudotsuga menziesii
McBride Spr., Mt. Shasta, Siskiyou Co.	74F55
Blodgett Forest, El Dorado Co.	80F114
Chionodes periculella (Bsk.)	Pseudotsuga menziesii
2 mi. W Anderson Springs, Lake Co.	80F21
1.5 mi. NE Angwin, Napa Co.	80E39
1 mi. SE Angwin, Napa Co.	80E44
Chionodes sabinianae Powell	Abies concolor
Tehachapi Mt. Park, Kern Co.	80F26.1
PLUTELLIDAE:	
Ypsolopha cervella (Wlsm.)	Pseudotsuga menziesii
8 km NE Cecilville, Siskiyou Co.	80F59.1
1 mi. W Angwin, Napa Co.	80E33
YPONOMEUTIDAE:	
Zelleria haimbachi Bsk.	Pseudotsuga macrocarpa
Mt. San Antonio, Manker Flat, San Bernardino Co.	81F106.1
Zelleria haimbachi Bsk.	Pseudotsuga menziesii
8 km NE Cecilville, Siskiyou Co.	80F59.2
TORTRICIDAE: OLETHREUTINAE:	
Epinotia radicana (Heinr.)	Abies concolor
McBride Spr., Mt. Shasta, Siskiyou Co.	70G2, 74F54.1, 80F87
Mill Crk., Warner Mts., Modoc Co.	70G15, 74G35.1
Fandango Pk., Warner Mts.,	74G31.1, G49.1

Modoc Co.	
Benton Mdws., Warner Mts., Modoc Co.	74G33.2, G51.1
Shield Crk., Warner Mts., Modoc Co.	74G39.1
9 air km NE Adin, Modoc Co.	80F100
8 air km NE Adin, Modoc Co.	80F103, G19.2
Soup Crk., Warner Mts., Modoc Co.	74G59.1
Hilton Spike Camp, Modoc Co.	80G21
Blodgett Forest, El Dorado Co.	79F29, F40
Epinotia radicana (Heinr.)	Abies magnifica
Mt. Shasta, Panther Mdws., Siskiyou Co.	70G3
Kaspian Rec. Area, Placer Co.	79G13
Epinotia radicana (Heinr.)	Pseudotsuga menziesii
5 mi. SW Ashland, Jackson Co., Ore.	70G9
10 air km SE Trinity Ctr., Trinity Co.	80F46
10 air km SW Gazelle, Siskiyou Co.	80F77.1
Blodgett Forest, El Dorado Co.	79F27, F37.1, 80F110
Plantation, Sonoma Co.	79F3.1
3 mi. N Branscomb, Mendocino Co.	80E54
1 mi. NE Cobb, Lake Co.	79E81, 80F14
1 mi. W Angwin, Napa Co.	79E60
Epinotia meritana Heinr.	Abies concolor
Manzanita Mt., Modoc Co.	74F77.1
Snell Spr., Warner Mts., Modoc Co.	74G29.1, 74G74.1
Tehachapi Mt. Park, Kern Co.	81F104.1
Epinotia trossulana (Wlsm.)	Abies concolor
Lily Lake, Modoc Co.	70F97
Scott Mt. Trinity Co.	80F55
9 mi. N Bartle, Siskiyou Co.	74F44.1
Mt. Shasta, Everitt Mem., Siskiyou Co.	74F59
Epinotia trossulana (Wlsm.)	Abies magnifica
Kaspian Rec. Area, Placer Co.	79G10.1

Epinotia trossulana (Wlsm.)	Pseudotsuga menziesii
Sea Ranch, nr. Airport, Sonoma Co.	85B24
TORTRICIDAE: TORTRICINAE:	
Acleris gloverana (Wlsm.)	Abies concolor
Fandango Pass, Warner Mts., Modoc Co.	70F25.1
Snell Spr., Warner Mts, Modoc Co.	74G65.1
9 air km NE Adin, Modoc Co.	80F95
8 air km NE Adin, Modoc Co.	80G19.1
Snowmans Summit, Mt. Shasta, Siskiyou Co.	80F93.1
Blodgett Forest, El Dorado Co.	79F40.1, 81F118
Acleris gloverana (Wlsm.)	Abies magnifica
Kaspian Rec. Area, Placer Co.	79G10.2, 81G13.1
Acleris gloverana (Wlsm.)	Pseudotsuga menziesii
4 km SW Etna, Siskiyou Co.	80F71
Blodgett Forest, El Dorado Co.	80G1
Plantation, Sonoma Co.	79F3
Archips argyospila (Wlk.)	Abies concolor
Angelus Oaks, San Bernardino Co.	82F25.1
Archips argyospila (Wlk.)	Pseudotsuga macrocarpa
Angelus Oaks, San Bernardino Co.	82F19
Archips argyospila (Wlk.)	Pseudotsuga menziesii
6 air km ESE Trinity Ctr., Trinity Co.	80F50
Mt. Shasta City, Siskiyou Co.	74F37
8 km NE Cecilville, Siskiyou Co.	80F59
10 air km SW Gazelle, Siskiyou Co.	80F77.2
Cazadero, Sonoma Co.	79F15
3 mi. N Longvale, Mendocino Co.	80F3
1 mi. NW Cobb, Lake Co.	80F18
2 mi. NE Angwin, Napa Co.	79E44.1
1 mi. W Angwin, Napa Co.	79E61, 80E34
Archepandemis coniferana Mut.	Abies concolor
Benton Mdws., Warner Mts., Modoc Co.	74F65.1, G15.2, G33.3

Mill Crk., Warner Mts., Modoc Co.	74F66.2
S. Deep Crk., Warner Mts., Modoc Co.	74F68.1, G19.2
Snell Spr., Warner Mts., Modoc Co.	74F79.2, G47.2
Fandango Pk., Warner Mts., Modoc Co.	74G13.1
Shields Cr., Warner Mts., Modoc Co.	74G21.1
Soup Cr., Warner Mts., Modoc Co.	74G23.2, G41.2
Argyrotaenia provana (Kft.) Abies concolor	
Snell Spr., Warner Mts., Modoc Co.	74G47.4
Manzanita Mt., Modoc Co.	74F78.1
Fandango Pk., Warner Mts., Modoc Co.	74G13.3
Shields Crk., Warner Mts., Modoc Co.	74G21.3
Mill Crk., Warner Mts., Modoc Co.	74G35.2
9 air km NE Adin, Modoc Co.	80F94
8 air km NE Adin, Modoc Co.	80G19.4
Donner Lake, Nevada Co.	79G3
Argyrotaenia provana (Kft.)	Abies magnifica
25 air mi E. Medford, Jackson Co., Ore.	70G13.1
Argyrotaenia provana (Kft.)	Pseudotsuga menziesii
10 air km SE Trinity Ctr., Trinity Co.	80F44
Blodgett Forest, El Dorado Co.	79F34
Argyrotaenia citrana (Fern.)	Pseudotsuga menziesii
1 mi. NE Occidental, Sonoma Co.	80E48
Argyrotaenia dorsalana (Dyar)	Abies concolor
S. Deep Crk., Warner Mts., Modoc Co.	74F68.2, G55.2
Shields Crk., Warner Mts., Modoc Co.	74F70.1, F71.2, G21.2, G39.2
Long Valley, Warner Mts., Lassen Co.	74F76.1, G43.1
Snell Spr., Warner Mts., Modoc Co.	74F79.1, 74G47.3
Fandango Pk., Warner Mts., Modoc Co.	74G13.2, G31.2, G49.2
8 air km NE Adin, Modoc Co.	80F105.2, G19.3
5 air km SW Etna, Siskiyou Co.	80F66
Argyrotaenia dorsalana (Dyar)	Abies magnifica
25 air mi. E Medford, Jackson Co., Ore.	70G13.2

Kaspian Rec. Area, Placer Co.	79G15
Argyrotaenia lautana Powell	Abies concolor
Tehachapi Mt. Park, Kern Co.	82F15.3
Angelus Oaks, San Bernardino Co.	82F25
Argyrotaenia lautana Powell	Pseudotsuga macrocarpa
Angelus Oaks, San Bernardino Co.	82G9
Manker Flat Cpgd., San Bernardino Co.	80F29.1
Choristoneura rosaceana (Harr.)	Pseudotsuga menziesii
1 mi. W Angwin, Napa Co.	80E31
Clepsis persicana (Fitch)	Abies concolor
Mt. Shasta, Siskiyou Co.	74F15
5 air km SW Etna, Siskiyou Co.	80F66.2
Sparganothis senecionana (Wlsm.)	Abies concolor
5 air km SW Etna, Siskiyou Co.	80F66.1
Sparganothis senecionana (Wlsm.)	Abies magnifica
2 mi. S Tahoe City, Ward Crk., Placer Co.	79G9.1
Kaspian Rec. Area, Placer Co.	79G10.3
Sparganothis senecionana (Wlsm.)	Pseudotsuga menziesii
5 air km SW Etna, Siskiyou Co.	80F68
Amorbia cuneana (Wlsm.)	Pseudotsuga menziesii
1.5 mi. N Branscomb, Mendocino Co.	80E57
PYRALIDAE:	
Dioryctria pseudotsugella Munr.	Abies concolor
Mill Crk., Warner Mts., Modoc Co.	74F66, G17, G35, G53
Cedar Mt., Warner Mts., Modoc Co.	74F67
S. Deep Crk., Warner Mts., Modoc Co.	74F68, F69, G19, G37, G55
Shields Crk., Warner Mts., Modoc Co.	74F70, G21, G39, G57
Soup Crk., Warner Mts., Modoc Co.	74F73, G23, G41, G59
Mahogany Ridge, Warner Mts., Modoc Co.	74F74

Manzanita Mt., Modoc Co.	74F77, F78, G27, G45, G63
Snell Spr., Warner Mts., Modoc Co.	74F79, G29, G47, G65
Fandango Pk., Warner Mts., Modoc Co.	74G13, G31, G49
Benton Mdws., Warner Mts., Modoc Co.	74G33, G51
Long Valley, Warner Mts., Lassen Co.	74F75, F76, G25, G43, G61
10 air km SE Trinity Ctr., Trinity Co.	80F48
Scott Mt., Trinity Co.	80F54
Tehachapi Mt. Park, Kern Co.	80F26, 81F101.1, 82F15.4
Dioryctria pseudotsugella Munr.	Pseudotsuga menziesii
10 air km SW Gazelle, Siskiyou Co.	80F77
11 air km SW MacDoel, Siskiyou Co.	80F82
1 mi. NW Cobb, Lake Co.	80F17
Angwin, Napa Co.	79E56
Robt. L. Stevenson Park, Napa Co.	8OF8
Manker Flat Cpgd., San Bernardino Co.	80F30
GEOMETRIDAE:	
Neoalcis californiaria (Pack.)	Pseudotsuga menziesii
2 mi. NE Angwin, Napa Co.	79E51
Angwin, Napa Co.	79E58
Cochisea sonomensis McD.	Pseudotsuga menziesii
8 km NE Cecilville, Siskiyou Co.	80F60
2 mi. NE Angwin, Napa Co.	79E48
Gabriola dyari Taylor	Pseudotsuga menziesii
1 mi. NW Cobb, Lake Co.	80F11
1.5 mi. NE Angwin, Napa Co.	80E38
Neptya umbrosaria nigrovenaria Pack.	Pseudotsuga menziesii
Alpine Lake, Marin Co.	78D24
Plantation, Sonoma Co.	79F5
Enypia packardata Taylor	Abies concolor
Tehachapi Mt. Park, Kern Co.	80F28

Enypia packardata Taylor	Pseudotsuga menziesii
2 mi. NE Angwin, Napa Co.	79E49
SATURNIIDAE:	
Hyalophora euryalus (Bdv.)	Pseudotsuga menziesii
Blodgett Forest, El Dorado Co.	79F32
1 mi. W Angwin, Napa Co.	79E72
ARCTIIDAE:	
Lophocampa argentata (Pack)	Pseudotsuga menziesii
Plantation, Sonoma Co.	79F9
1 mi. NW Cobb, Lake Co.	79E82
2 mi. NE Angwin, Napa Co.	79E47
Angwin, Napa Co.	79E57
NOCTUIDAE:	
Syngrapha celsa (H. Edw.)	Abies concolor
Benton Mdw., Warner Mts. Modoc Co.	74G15.1, G33.1
Soup Cr., Warner Mts., Modoc Co.	74G23.1
Achytonix epipaschia (Grote)	Pseudotsuga menziesii
10 air km SE Trinity Ctr., Trinity Co.	80F45
Anomogyna mustelina (Smith)	Abies concolor
2 mi. NW Fandango Pass, Modoc Co.	74G13
Egira crucialis (Harv.)	Pseudotsuga menziesii
1 mi. W Angwin, Napa Co.	79E62, E64
Egira simplex	Pseudotsuga menziesii
1 mi. W Angwin, Napa Co.	79E73
Feralia deceptiva McD.	Pseudotsuga menziesii
Blodgett Forest, El Dorado Co.	79F35

Appendix 6.2

Key to late instar, foliage-feeding lepidopterous larvae on *Abies* and *Pseudotsuga* in California.*

The key includes four Geometridae that are reported as associates of *Choristoneura* in the Pacific Northwest and Rocky Mountain states (Furniss and Carolin 1977, Stevens *et al.* 1984), which are known to occur in California but were not detected during our larval surveys and are not characterized in the text: *Lambdina fiscellaria* (Guenee), *Melanolophia imitata* (Walker), *Nematocampa limbata* (Haworth), and *Nepytia phantasmaria* (Strecker).

1. Abdominal prolegs present on more than 5 segments; a single large 'ocellus' on each side of the head . . . . Hymenoptera (sawflies)
   Abdominal prolegs present on 5 or fewer segments, including anal; 6 tiny stemmata ('ocelli') around a darkened area on each side of HC (Lepidoptera) 2
2. Abdominal prolegs present on segments 3-6 inclusive, and anal 13
   Abdominal prolegs lacking from segments 3 and 4 3
3. Abdominal prolegs present on segment 5 . . . . *Syngrapha celsa*
   Abdominal prolegs absent from segment 5 (Geometridae) 4
4. Abdominal segments 2 and 3 with long, paired, slender projections on dorsum . . . . *Nematocampa limbata*
   Abdomen without such projections 5
5. Body generally green with pale longitudinal stripes 6
   Body generally gray, tan, brown with darker markings 9
6. Head reddish . . . . *Enypia packardata*
   Head green 7
7. Stout-bodied with a thin reddish D line, broad DL stripes and diagonal lines on each segment dorsally that render a diamondback pattern . . . . *Cochisea sonomensis*
   Typically slender loopers without diagonal lines dorsally 8
8. Head without conspicuous spots; body dark foliage-green with thin, pale D and DL lines not edged darker . . . . *Melanolophia imitata*
   Head pale green with conspicuous black spots; body with no D line, DL lines broad, white, distinctly edged with dark green . . . . *Nepytia phantasmaria*
9. Head and body generally pale gray; a broken yellow D line, white DL stripe edged with black, and longitudinal black streaks on L, SV . . . . *Lambdina fiscellaria*
   Head and body generally tan or brown 10
10. Mature larva small (length 22-24 mm), slender larva, pale reddish or yellowish brown with darker red-brown D, DL lines and yellow L . . . . *Eupithecia annulata*
    Larger (25-35 mm) and generally darker larvae 11

* HC=head capsule; A=abdomen; D=dorsal; DL=dorsolateral; L=lateral; SV=subventral

11. Brown with darker markings, without pale stripes; hold themselves erect from substrate to resemble twigs . . . . *Neoalcis californiaria*
    Pale to reddish brown, with longitudinal paler bands; not stick-like in appearance and behavior 12
12. Pale tan to reddish brown with poorly defined longitudinal markings, a pale D line, and brownish black to orange laterally . . . . *Gabriola dyari*
    Pale cream-colored dorsally, reddish brown laterally with broken orange D and DL stripes edged with black . . . . *Nepytia umbrosaria*
13. Abdominal crotchets arranged in a series along the mesal half of each proleg 14
    Abdominal crotchets arranged in a circle or nearly complete ellipse 20
14. Body with elongate, dense hairs 15
    Body without conspicuous hairs 16
15. Mesothorax and abdominal segments 1-3 with dense tufts of pale hair, toothbrush-like . . . . *Orgyia pseudotsugata*
    Hair of body more or less uniform in density, no toothbrush-like tufts . . . . *Lophocampa argentata*
16. Large, robust, green larva with paired, prominent yellow dorsal tubercles, those of the thorax banded with black . . . . *Hyalophora euryalus*
    Smaller larvae without paired dorsal tubercles 17
17. Head whitish with reddish markings or entirely red 18
    Head green 19
18. Head unmarked red or chestnut colored; body olive-gray dorsally with pale, D, SD, and SV lines and venter pale . . . . *Egira* (*crucialis*, *simplex*)
    Head pale with brownish to red markings; body gray with a white D line edged by black, broad purplish L and yellow SV stripes . . . . *Anomogyna mustelina*
19. A yellowish stripe through the spiracles; 8th segment slightly broadened and round-humped . . . . *Feralia deceptiva*
    A dark line through the spiracles; 8th segment not humped . . . . *Achytonix epipaschia*
20. Prothoracic L group with 2 setae; DL1 setal bases of mesothorax and 8th abdominal segment each with a sclerotized ring . . . . *Dioryctria pseudotsugella*
    Prothoracic L group with 3 setae; DL1 setae without sclerotized rings 21
21. Setae L1 and L2 on A1-8 on separate pinacula with L2 displaced posteroventral of spiracle 22
    Setae L1 and L2 on A1-8 close together, on the same pinaculum, ventral oranteroventral of spiracle 23
22. Larva tan with broad, dull orange DL stripes. Abdominal prolegs normal, shorter than thoracic legs . . . . *Zelleria haimbachi*
    Larva green. Abdominal prolegs elongate, as long as thoracic legs; setal pinacula conspicuously black . . . . *Ypsolopha cervella*

23. Seta SD1 on A8 dorsal or dorsoanterior of spiracle; A9 with setae D2 usually on separate pinacula and farther apart than each is from its corresponding D1 (Gelechiidae) 24
    Seta SD1 on A8 anterior or anterodorsal of spiracle; A9 with setae D2 usually on a single mid-dorsal pinaculum and closer together than each is to its corresponding D1 (Tortricidae) 27
24. Body more or less unicolorous, pink to yellowish brown or green; pinacula small but dark and conspicuous 26
    Body marked with irregular linear pattern of white and pink to dark red-brown . . 25
25. Body dark reddish brown with irregular, parallel DL and L whitish stripes . . . . *Chionodes abella* and *periculella*
    Body mostly pale, with irregular DL and L stripes of reddish . . *Chionodes sabinianae*
26. Body uniform green, becoming pink-tinged dorsally at maturity, setae of anal plate disproportionately elongate, black, conspicuous . . . . *Telphusa sedulitella*
    Body pink to yellowish brown with dark thoracic and anal shields; setal pinacula small, dark, and conspicuous but anal setae not disproportionately elongate or conspicuous . . . . *Coleotechnites* species
27. Mature larvae in needle mines; head and thoracic shield dark brown, body cream-colored to yellow-green . . . . *Epinotia meritana* (and *E. trossulana*)
    Late instars external feeders; if body pale, thoracic shield not dark brown 28
28. Body tan to rust-brown or dark brown 29
    Body green, yellow-green, or greenish gray 34
29. Body (including thoracic shield) tan with diffuse rust-colored D and DL bands; pinacula not strongly differentiated . . . . *Epinotia radicana*
    Thoracic dorsum dark brown to rust, D and DL bands dark if discernible and/or thoracic shield differentiated, with black markings; pinacula often conspicuously pale 30
30. Body tan to brownish, head orange, thoracic shield orange with a strong black posterior rim . . . . *Zeiraphera* species
    Body brown to rust with darker HC and DL bands; thoracic shield may have dark markings but not a black posterior rim 31
31. Body tan with irregular chocolate brown DL bands; thoracic shield tan with a black spot posterolaterally on each side . . . . *Syndemis afflictana*
    Body primarily dark brown or pale with rust bands; thoracic shield mostly blackish (penultimate), rust or dark brown 32
32. Body primarily dark chocolate brown, with pinacula conspicuously paler . . . . *Archepandemis coniferana*
    Body brick red to cream-colored with rust-brown bands, with pinacula conspicuously paler 33

33. Body primarily brick red with pale DL bands; cismontane California in association with *Pseudotsuga*. . . . . *Choristoneura carnana*
Body mostly pale with D and DL rust-brown areas; higher elevations and east of the Sierra crest, primarily in association with *Pinus contorta murrayana*. . . . . *Choristoneura lambertiana subretinana*
34. Head and thoracic shield tan, weakly pigmented, appearing green, similar to body color 35
Head and thoracic shield orange-brown to black, distinctly darker than body. . . . 37
35. Larva pale yellow-green, head yellow-green with dark areas laterally on cranial lobes . . . . *Clepis persicana forbesi*
Larva pea green, head concolorous, without dark lateral markings. 36
36. Setal pinacula yellowish, paler than body color, not elongated; abdominal spiracles of segments 1-7 larger than setal bases . . . . *Choristoneura retiniana*
Setal pinacula scarcely paler than body, elongated posteriorly; abdominal spiracles small, about the size of setal bases . . . . *Argyrotaenia* species**
37. Abdominal setae SD1 and SD2 on the same pinaculum; D and SD pinacula distinctly paler than body 38
Abdominal setae SD1 and SD2 of 1-8 on separate pinacula; D and SD pinacula not paler than surrounding integument . . . . *Acleris gloverana*
38. D and often SD pinacula of mesothorax elongated posteriorly; dorsum of body not contrastingly darker than laterally 39
D and SD pinacula of mesothorax more or less round; dorsum of body contrastingly dark olive green . . . . *Archips argyrospila*
39. Spiracles on abdominal segments 1-7 much larger than a seta base; prothoracic shield tan with small anterolateral black spots . . . . *Choristoneura rosaceana*
Spiracles on abdominal segments 1-7 small, about the size of a seta base; prothoracic shield entirely dark or orange-brown with black lateral rim. 40
40. Larger larva (length ca 22-26 mm) with orange-brown HC and thoracic shield, both marked by black laterally . . . . *Amorbia cuneana*
Smaller larva (length ca 16-21 mm) with dark brown to black HC and shield, not distinctly marked with black laterally . . . . *Sparganothis senecionana*

**Larvae of the species of *Argyrotaenia* discussed here cannot be distinguished superficially; 3 are allopatric; *A. citrana* is coastal on *Pseudotsuga*, *A. dorsalana* occurs in the Mt. Shasta–Modoc region and Sierra Nevada in association with *Abies*, while *A. lautana* is known in the Greenhorn Mountains, Tehachapi, and Transverse Ranges of southern California. *A. provana* is sympatric with *A. dorsalana*, and has yellowish DL and L stripes, but these are lost in the final instar according to Stevens *et al.* (1984).

7

Evolutionary Interpretation, Taxonomy, and Nomenclature

Jerry A. Powell
Essig Museum of Entomology
University of California, Berkeley, California

John A. De Benedictis
Department of Entomology
University of California, Davis, California

Abstract

With a team effort using several approaches, we have been able to define *Choristoneura* species groups by a three-tiered analysis—reproductive isolation through intraspecific recognition (pheromone chemistry and correlated behavior), ecological separation (larval host trees), and morphological distinction (adult, larval)—enabling us to reach conclusions that the traditional approach, study of museum specimens, could not. Because relationships among the three species groups are not static over the western Nearctic, however, we are unable to propose taxa (species, subspecies) that would reflect relationships completely. Two entities that would be defined as species by any criteria are sympatric at middle elevations in California (*Choristoneura retiniana* and *C. carnana*) and in the Great Basin Ranges (*C. retiniana* and *C. occidentalis*). North of about 42° N lat., however, the reproductive isolation between these species breaks down in eastern Oregon and possibly southern Idaho.

We used fossil records of floristic changes in Nearctic conifer forests to reconstruct a scenario of the evolutionary history of western *Choristoneura.* The sympatry of *C. retiniana* and *C. carnana* in California appears to be relatively recent, with *C. carnana* moving into the area with *Pseudotsuga* from the south, *C. occidentalis* on *Pseudotsuga* from the north, and *C. retiniana* migrating with *Abies concolor* from the southeast during the late Pliocene or Pleistocene, as the Sierra Nevada uplift occurred. The hybrid zone in Oregon appears to be of quite recent origin, perhaps 7,000 yr BP, in the area of latest contact of the Rocky Mountain and coastal populations of the conifers following separation by glacial advance during late Wisconsin times.

We provide a synopsis of formal taxa, with diagnoses of adult characters and photos of exemplar specimens. Nonetheless, identification based on single specimens is impossible for many populations.

We regard *Choristoneura retiniana* (Walsingham) s. str. (Type Locality: Mt. Shasta, California) to be represented by populations in the southern Cascade Range and Warner Mountains of California and southeastern Oregon, which are highly variable, probably the result of genetic introgression from the hybrid populations to the north and west; we apply the name *C. spaldingiana* Obraztsov (TL: Provo, Utah) in a subspecific sense **(new status)** to Great Basin and Sierra Nevada populations of *C. retiniana*, which are dimorphic in both sexes but exhibit less individual variation, with the dimorphism geographically polytypic.

Choristoneura carnana (Barnes and Busck) s. str. (TL: Mt. Baldy, California) is restricted to the Transverse Ranges in California, associated with *Pseudotsuga macrocarpa*; other populations of this species feed on *P. menziesii*; the name *californica* Powell (TL: Anderson Springs, Napa Co., California) can be applied to North Coast Ranges populations, while those of the Sierra Nevada have no trinomial designation.

To the north, populations west of the Cascades in Oregon, in Washington, the Great Basin ranges, and Rocky Mountains are primarily associated with *P. menziesii* and are subjectively grouped as *C. occidentalis* Freeman (TL: Klickitat Co., Washington). Preliminary data from morphology and cross attraction by females from Satus Pass, which is near the type locality, to males of Sierran *C. carnana* suggests that the type locality is north of the hybrid zone.

Choristoneura lambertiana (Busck) s. str. (TL: Ashland, Oregon) is restricted to the Siskiyou Range in southern Oregon and northern California, associated with Sugar Pine, *Pinus lambertiana*; we apply the name *subretiniana* Obraztsov (TL: Monachee Meadows, Tulare Co., California) broadly to populations of the eastern Sierra Nevada and eastern Oregon that are associated with *P. contorta*, despite considerable polytypy in adult color pattern from south to north; populations of the eastern Rocky Mountains living on *P. ponderosa* can be assigned the subspecific name *ponderosana* Obraztsov (TL: Sugarloaf, Boulder Co., Colorado), while those of the northern Rocky Mountains and Great Basin ranges exhibit intermediate aspects of color pattern and are not given trinomial designation.

We propose no new taxa or new synonymies.

INTRODUCTION

The foregoing series of studies in this volume demonstrates that there are three species-group lineages of conifer-feeding *Choristoneura* in the western United States: one that feeds on *Abies*, one primarily on *Pseudotsuga*, and one primarily on *Pinus*. In the northern Sierra Nevada, California, and in southern Utah, adults of all three fly together. In these situations and throughout the California-Nevada-Utah region at any site where two of the species are sympatric, premating isolation maintains the integrity of each. Males are attracted to specific sex pheromones of their corresponding females and not to those of other members of the complex. The chemical cues remain constant within each species across broad geographic regions of the southwestern Nearctic, even though morphological and color pattern characteristics of the moths vary considerably among conspecific populations.

As systematists, we have two goals in research: first, to understand the evolutionary or biological (genetic) relationships among the animals; and second, to develop a taxonomy that reflects that knowledge. If we achieve these objectives, names will be assigned to the taxa that facilitate communication among biologists, and the classification will enable biologists to make meaningful predictions based on natural relationships.

If the biological relationships in *Choristoneura* summarized above were consistent throughout the western Nearctic, we would have achieved an ideal solution to a difficult taxonomic problem caused by morphological similarity of the species. Differences are subtle and variable, more so in some characters among populations within a species than between species. With a team effort using several approaches, we have been able to define species concepts on a three-tiered assessment: reproductive isolation (pheromone chemistry and correlated behavior), ecological separation (larval hostplants), and morphological distinction (adult, larval). This combined assessment permitted us to draw conclusions that the traditional approach, examining museum specimens, could not.

Unfortunately, relationships among the three species groups are not static over the whole array of conifer-feeding *Choristoneura* in the western United States. Two entities that would be defined as species by any criteria are sympatric at middle elevations throughout much of California: *C. retiniana* (Walsingham) is dimorphic in both sexes, has green larvae that feed on *Abies*, and produces a pheromone with E-ll-tetradecenyl acetate as its principal component; *C. carnana* (Barnes and Busck) is monomorphic, has red-brown larvae that feed on *Pseudotsuga*, and produces a pheromone with E-11-tetradecenyl aldehyde as its principal component. Adults in sympatric populations of the two fly synchronously and are 100% reproductively isolated by males' response to pheromones. The same set of relationships exists between *C. retiniana* and *C. occidentalis* Freeman in the Great Basin ranges, and comparable distinction prevails between any pair of the three species where pine-feeding populations of *C. lambertiana* (Busck) are sympatric with the Abietoideae-feeders.

As is documented in detail in Part 2 (Powell and De Benedictis 1995), however, at about 42° N Lat., the northern border of California, Nevada, and Utah, the reproductive isolation of *Abies*- and *Pseudotsuga*-feeding lineages (and possibly *Pinus*-feeding as well) breaks down. Most or all populations hybridize, morphological characteristics of adults and larvae blend, and the pheromones that are selected by one species exclusively in more southerly localities attract a similarly heterogeneous array of moths in Oregon east of the Cascade Range (ibid, Volney *et al.* 1984).

We have made considerable progress in understanding the natural relationships of western spruce budworms, but as taxonomists we have to delimit taxa and apply the binomial or trinomial system of nomenclature. This does not provide much flexibility to accommodate of situations of evolutionary intermediacy, degrees of similarity among allopatric populations, and so on. An animal's name represents its status as a species; and if subspecies concepts are invoked, each population can be assigned to a trinomial subjectively, but degrees of relationship among them cannot be expressed by the names. Hence, we are faced with a paradox: if *C. retiniana* and *C. carnana* in California are treated as species, and in every respect they qualify, we then must admit that they do not meet traditional criteria of species north of the Oregon border. If we decide the whole Abietoideae-feeding complex is a single species, we are left to explain how this species can have two sympatric, reproductively isolated components in California. Moreover, the polytypic pine-feeding populations have been assigned to one species, *C. lambertiana*, on the basis of larval host and pheromone chemistry, but we have no direct evidence that distant populations, such as *C. l. ponderosana* in Colorado and *C. l. subretiniana* in the Sierra Nevada, would hybridize if in contact. The allozyme data (Harvey 1994) suggest that *C. l. ponderosana* populations of the eastern slope of the Rocky Mountains are genetically distinct from *C. l. subretiniana* in eastern Oregon at a level comparable to that which either shares with other western species of conifer-feeding *Choristoneura*. Furthermore, we are not sure that *C. lambertiana* maintains its distinctness from the Abietoideae-feeding lineage in parts of Oregon. If we interpret subspecies more broadly than as local, isolated populations, it is arbitrary where many populations, such as those of the northern Rocky Mountains, are assigned.

In the discussion that follows, we speculate on the evolution of the complex relationships among these conifer-feeding *Choristoneura*. We outline a classification and nomenclature that attempts to reflect this knowledge (Table 7.1) —but because these insects do not conform to our traditional biological concepts, the classification cannot represent all of the relationships accurately.

HISTORY OF NEARCTIC CONIFEROUS FORESTS DURING THE TERTIARY

Speculation concerning the evolution of species relationships in Lepidoptera is dependent upon interpretation of indirect evidence, because there is no fossil record of intrageneric events in moths. For phytophagous taxa that are host specific, such as the spruce budworm species, fossil records of floristic changes may provide the best data for inference of origins. In western North America, however, the later Tertiary and Quaternary history is difficult to interpret because of the wide range of geological events and resultant geographical and climatic diversity. There are only scattered macrofossil floras for any one epoch, and well dated sites of appropriate ages with fossil pollen are few. Moreover, lack of suitable pollen sites in ecotonal areas that are most sensitive to climatic changes (e.g., mountain slopes), and the ease with which pollen is blown from one vegetation zone to another, hinder attempts at summarizing past events (e.g., Axelrod 1976, Porter 1983, Tsukada 1982, Wright 1983).

We use present day geographic names as points of reference in the following discussion. Major features of evolution in forest composition in the western United States have been inferred by Axelrod (1976). His hypothesis reconstructs a land of low to moderate topography that supported a rich mesic conifer-hardwood forest in the early Tertiary, gradually evolving to the highly complex land form of today. Generally drier climatic regimes replaced more humid ones; superimposed on the geographic diversity, this resulted in zonal evolution of specialized, relatively depauperate forest types.

Paleogene Conifer Forests

Fossil records of conifers that are equivalents of modern species in the western Nearctic extend back to the mid-Eocene. They occurred in subhumid forests of the interior Cordilleran axis (present day Rocky Mountains) and included *Abies, Larix, Picea, Pseudotsuga, Sequoia, Taxus*, and numerous hardwoods, many of which are related to descendants that live in deciduous forests of the eastern United States or subtropical areas of abundant summer rainfall. Later, as the continental trend to decreased precipitation spread over the interior, mesic taxa gradually were confined coastward (Axelrod 1976).

A drier climate had spread over the southern Rocky Mountain region by mid-Oligocene (32 million years before the present: 32 m yr BP). The flora of southwestern Colorado had a conifer dominated forest, with *Abies, Picea, Pseudotsuga*, and associated deciduous shrubs, in Oligo-Miocene transitional times (27 m yr). Thus, a nearly modern Rocky Mountain conifer forest appeared in that area long before one entered the Sierran axis—middle and late Miocene floras from areas of the present day Sierra Nevada were dominated by subtropical deciduous hardwoods. Hence, equivalents of *Choristoneura occidentalis* and/or *C. lambertiana* may have existed in the interior for 10 to 20 m yr before they appeared in the Sierra Nevada.

During the early Tertiary, the California-Oregon region was still a humid subtropical area. Much of the present coastal plain was submerged; there were low volcanic mountains along the Cascade chain, but the major uplifts of the west coast mountains had not yet taken place. Rich mesophytic forests dominated by *Metasequoia* occurred broadly over this plain and included many hardwood equivalents of species found in eastern American and tropical forests today, e.g., *Juglans*, *Ficus*, and *Persea*. By late Oligocene, rich conifer-hardwood forests with *Abies, Taxus,* and *Tsuga* and many deciduous trees had appeared in east-central Oregon.

Neogene Conifer Forests

At the beginning of the Miocene, spruce-fir forests had not yet appeared in the coastal lowlands of the Pacific Northwest, where they now dominate. The interior forests, in the Great Basin and northward, began to segregate into northern and southern types during the Miocene. Those of the north (eastern Oregon, Idaho, north-central Nevada) had more mesic species—*Sequoia*, *Thuja*, *Tsuga*, and *Pseudotsuga sonomensis* (the equivalent of modern *P. menziesii*), and numerous hardwoods of mesic requirements; while floras of the south (eastern and south-central Nevada, southern Idaho) were characterized by *Abies concoloroides* (the precursor of *A. concolor*), *Pinus ponderosa*, and *Sequoiadendron*, with fewer mesic hardwoods and more oak-woodland and chaparral taxa (Fig. 7.1). Early to mid-Miocene (25 to 15 m yr BP) forests of the Great Basin and border areas were very rich compared to modern equivalents, with 8 to 12 species of conifers (Axelrod 1976).

In the Miocene, global temperatures were warmer in the northern hemisphere than today, resulting in more uniform, warmer, subtropical-like climates, characterized by summer rains. The Cascades and Sierra Nevada were still relatively low elevation; their major uplifts occurred in the Pliocene and Pleistocene. Hence there was a broad marine influence and more homogeneous vegetation in northwestern North America. Ocean temperatures shifted after mid-Miocene (15-16 m yr), causing temperature ranges of greater amplitude in the interior. This trend was accompanied by uplift of the mountains, accentuating the more extreme temperatures and aridity; early Pliocene conifer forests of the interior were restricted to higher elevations, and island-like aspects of western forests began to take form. By early Pliocene (8-10 m yr), rainfall, especially summer rain, is inferred to have decreased in the lowlands east of the Sierra Nevada to 20-25 inches annually from 30-35 inches in late Miocene (12-13 m yr). Forests of *Sequoiadendron* disappeared, and were replaced by oak-juniper woodland (Axelrod 1976).

As the progressive confinement northward and coastward of mesic forests continued, the residual floras of the Great Basin took on modern aspects. Mio-Pliocene transitional and early Pliocene floras of western Nevada had *Abies concoloroides* and *Picea* but no *Pseudotsuga* (Chaney 1944, Axelrod 1976). *A. concoloroides* occurred in forests characterized by broadleaf hardwoods (*Lithocarpus, Arbutus*, etc.) in addition to *Sequoiadendron* and other conifers (Fig. 7.2). In contrast, *Abies concolor* now lives in the middle to upper part of the conifer belt, for the most part well above the broadleaf taxa. Available evidence suggests that this restriction to higher levels occurred later, perhaps not until the Pleistocene. Axelrod presumed the summer climate at lower elevations became too hot and dry for seedling establishment by *A. concolor*.

Hence it is not possible to define precisely when *A. concolor* and *Pseudotsuga menziesii* became ecologically separated to the degree they are today in California, but the record suggests that this may have begun with their precursors in the Pliocene, with increasing aridity forcing *Pseudotsuga* coastward, and increasing summer temperatures forcing *A. concoloroides* into higher elevations (Fig. 7.2). When this occurred, it may have set the stage for differentiation of *Choristoneura retiniana* in Great Basin populations in comparable situations to those of the Tehachapi and Clark Mountains today, where *C. retiniana* occurs and other Abietoideae-feeding *Choristoneura* do not. Character traits such as the cryptic green larva and pupa and the color dimorphism in the adults, as well as more subtle morphological features, narrower forewing, fewer basitarsal spines, and broader uncus tip (De Benedictis Part 4 this vol.), may have been established in isolated populations which were derivatives of either a *C. occidentalis*-like or *C. lambertiana*-like ancestor in the Rocky Mountains.

Pacific Coast Pinaceae Forests

The fossil record of forest types in the Oregon-California area is fragmentary throughout the Neogene. Most described floras come from lake beds in lowland areas. Records of transitional montane forests were less likely to have been preserved and/or they were eliminated subsequently by uplifts and erosion from the steep gradients or by glaciation. As a result, there is a comparatively complete picture of Pliocene floras in central coastal lowland (Axelrod 1944), but the few records for the Sierra Nevada are primarily floras from the lower slopes, even in late Pliocene and early Pleistocene times (Axelrod and Ting 1960, 1961, Axelrod 1976).

During the Miocene (17-23 m yr floras) the Sierra Nevada region had low topographic relief and supported a deciduous hardwood forest. The west slope foothills had pine-oak woods and a modern coastal type forest, with *Sequoia* and equivalents of *Arbutus*, *Cercis*, *Cornus*, and *Umbellularia* (Condit 1944, Axelrod 1976) but neither *Pseudotsuga* nor *Abies*. The Denton Flora (near present Blairsden, Plumas Co., Calif.) had a similar composition. This area is believed to have risen 1,500-1,800 m since the Pliocene.

Therefore, we do not know when modern conifers appeared in the Sierra Nevada or North Coast region of California. Only 13 of 30 Pliocene floras in Oregon, California, and western Nevada have any Pinaceae. In California, only 3 of 21 floras have Pinaceae other than *Pinus*: *Abies premacrocarpa* in mid-Pliocene at Mt. Eden in southern California, and *Pseudotsuga sonomensis* in late Pliocene in the Santa Clara and Sonoma floras. The last had a forest like that of the modern North Coast, with *Abies* (but not *concoloroides*), *Picea*, *Tsuga*, and *Pseudotsuga* (Chaney 1944, Axelrod 1944, 1976) (Fig. 7.2).

Thus *Pseudotsuga* existed in coastal California, and *P. premacrocarpa* had differentiated from *P. sonomensis* during the Pliocene as early as 4-6 million years ago. Presumably this occurred in absence of other Abietoideae. The spread of Douglas-fir into California may have provided the isolated, marginal situation for differentiation of *Choristoneura carnana* from an ancestor associated with *Pseudotsuga* of the interior. The phenetic similarity of populations in Coahuila, Mexico, and northern Arizona to *C. carnana*

carnana suggests a more southerly origin of *C. carnana* (De Benedictis Part 4 this vol.). The clinal change in *Choristoneura* morphology that we see along the North Coast Range-Cascades axis may represent south to north differentiation, or alternatively, recent recontact after separation of northern and southern populations during the Pleistocene.

Quaternary Conifer Forests

At the beginning of the Pleistocene, increased precipitation presaged the buildup of the major ice sheets, and there were rich mixed conifer forests in the low to moderate elevations of the southern Sierra Nevada (Axelrod and Ting 1960, 1961). Pollen floras indicate forested regions characterized by several species of pines, *Pseudotsuga*, *Tsuga*, and *Abies* (*concolor* and other species) in the present Mammoth Lakes basin, Kern Plateau, and Owens Valley. Eastward, there were only pines in the Panamint Valley. Hence it is probable that *A. concolor* existed continuously at moderate elevation, in islands much larger than now, across the southern Great Basin and the present northern Mojave into southern California (Axelrod 1976).

It may be inferred that *Abies concolor* or its equivalent became sympatric with *Pseudotsuga macrocarpa* in southern California (Axelrod 1966) and *P. menziesii* in the southern Sierra Nevada (Axelrod and Ting 1960, 1961) beginning in late Pliocene or early Pleistocene times. This scenario provides the possibility of *Choristoneura retiniana* expanding its range from the Great Basin and establishing sympatry with *C. carnana* in southern California after several million years isolation of *C. retiniana* from the Rocky Mountain population of *C. occidentalis*. Thus there has been a long period (2-3 m yr) in which *C. carnana/occidentalis* and *C. retiniana* could have stabilized as sympatric, reproductively isolated species.

Floristic events in the northern California-Oregon conifer forests are also poorly known through the Pleistocene. Analysis of radiometric data has enabled a more realistic appraisal of the chronology of glaciation. There were many fluctuations in the spreading and shrinking of the ice sheets, superimposed upon a general downward trend of secular temperatures that began much earlier than the Pleistocene (Flint 1971). As global temperatures declined, there was widespread uplifting which continued through the Pliocene into later times, and temperatures on newly elevated lands were lowered, favoring spread of snow and ice.

Reconstructions of the extent of mountain glaciations in the western United States are less precise than those for the fluctuations of the southern margins of the major ice sheets (Porter 1983). Stratigraphic relationships suggest that the culminating advances of alpine glaciers were broadly in phase with those of the northern ice sheets. The northern California-Oregon region was much more strongly affected by Pleistocene glaciation than were southern California areas where *Pseudotsuga-Abies* forests occurred. Volney (1985) has reviewed the Holocene biogeography of North American conifer forests, with emphasis on relationships of hosts of *Choristoneura fumiferana* and *C. occidentalis*; our interpretation of events leading to present day distributions of the other western species benefited from discussions with Volney.

During glaciation in late Wisconsin time (± 16,000-10,000 yr BP), spruce and Douglas-fir forests, and presumably *Choristoneura* populations, were split into three components (Fig. 7.7). Conifer forests of eastern North America were shifted southward and compressed into a narrow band of only 2-5° latitude, roughly from Virginia and the Carolinas westward to Missouri and northern Mississippi, 22,000-16,000 yr BP. *Picea* forests were present across the Great Plains during late Pleistocene but gave way to treeless prairie ca. 11,000 yr ago (Porter *et al.* 1983, Péwé 1983). Migration northward and westward from this refugium was determined by glacial retreat, and forests of 20% *Picea* reached the southern edge of the Great Lakes by ca. 11,000 yr BP, the west end of Lake Michigan by 10,000 yr ago. Subsequent westward spread probably reestablished contact with refugia of the Rocky Mountain foothills in the northern prairie region, and with the white spruce forest that spread eastward from the Alaskan refugium, by ca. 8,000-9,000 yr BP. The boreal forest reached its approximate northern limits about 6,000 yr BP, and the western limits of its range 2,000-3,000 yr ago (Porter *et al.* 1983). Thus it is likely that eastern populations of spruce budworm were contiguous across the northern plains to the foothills of the Rocky Mountains during parts of the Pleistocene, were isolated by increasing aridity and elimination of *Picea* from the plains, and were reconnected with Rocky Mountain populations after a period of only 3,000-4,000 yr (ca. 11,000 to 8,000 yr BP).

The situation in the west is more complex, owing to the extreme topographic diversity, with mountain ranges oriented in north-south axes across the front of influence caused by glacial advance, because the fossil record is even more fragmentary, and because more host tree species are involved. The Cordilleran Ice Sheet extended from central Alaska nearly to the Columbia River, where it attained maximum advancement ca. 14,000 yr BP. Eastward, it met the Laurentide Ice Sheet at the Rocky Mountains near the present Canadian border. Southward, along the Cascades-Sierran axis and in the Basin Ranges and northern Rocky Mountains, alpine glaciers formed, the extent and fluctuations of which are poorly documented (Crandell 1965, Denton and Hughes 1981, Porter *et al.* 1983, Wahrhaftig and Birman 1965) (Fig. 7.7). During periods of maximum extent, these montane glaciers extended down to 2,000 ft. (610 m) in the northern Cascades; to 8,000 ft. (2,240 m) in the Sierra Nevada; and to somewhat higher elevation at comparable latitudes of the interior mountain ranges. In each case periglacial zones of permafrost formed, characterized by tundra-like vegetation, with perhaps scattered *Pinus albicaulis*, but no Abietoideae, above the spruce-fir forests (Péwé 1983).

There were broad barriers to conifer forests during maximum glaciation, probably considerably longer than just the Wisconsin maximum. Glaciation was particularly influential in the Rocky Mountains, where much broader expanses of high elevation country existed, and a wide zone of tundra-like habitat spread from the southern margin of the Cordilleran Ice Sheet southeastward into Colorado (Fig. 7.7) (Porter 1983, Péwé 1983).

Forests of the western United States survived in refugia at lower elevations and in some places well south of their present ranges. Evidence from macrofossils and wood rat middens in the southwest places spruce-fir forests about 280 miles (450 km) south and

3,000 ft. (915 m) downslope from their modern counterparts (Baker 1983, Cole 1982, Martin and Mehringer 1965). A major consequence of the alpine glaciation in the west was that Douglas-fir populations were split into two portions, one on the Pacific Coast and the other on the lower slopes of the Rocky Mountains, corresponding to the two varieties now recognized nomenclaturally by botanists. The Rocky Mountain *Pseudotsuga menziesii* var. *glauca* (Little 1953) is a more boreal race, ranging to 11,000 ft. (3,350 m), above *Pinus ponderosa* in the southern Rocky Mountains, and to 8,000 ft. (2,450 m) at 45° N Lat., while the typical coastal race occurs below 4,000 ft. (1,250 m) at a comparable latitude (Silen 1978). The northern limits of refugia of Douglas-fir in the Basin Ranges and Rocky Mountains have not been documented, but pollen analysis indicates that the northern and eastward spread reached southern Idaho by 11,000 yr BP, and advance into northern Idaho and reestablishment of contact with the Pacific Coast race at the eastern edge of the Cascades occurred about 7,000 yr BP (Baker 1983, Silen 1978, Tsukada 1982).

EVOLUTIONARY ORIGIN OF CONIFER-FEEDING *CHORISTONEURA* IN THE WESTERN UNITED STATES

Recent data on genetic relationships among *Choristoneura* based on isozyme analysis (Harvey 1994) and on Mitochondrial DNA (Sperling and Hickey 1994) contradict species-group proposals based on pheromone types (Sanders 1971), host trees (Powell 1980), or aedeagal spurs (Dang 1985) and do not support a species relationships hypothesis based on haemolymph polymorphism (Harvey and Stehr 1967). In contrast to dichotomies based on other lines of evidence, both Harvey and Sperling found that *C. fumiferana* is the most distinct species.

Genetic distances were significant but small for isozyme heterogeneity (maximum Nei = 0.232), with most genetic distances coefficients between species pairs well below what might be expected at the species level. Among western Nearctic species, the genetic distances were 0.001-0.085, markedly less than those usually encountered between congeneric species in other animals, including *Drosophila* (Ayala *et al.* 1974) and most vertebrates other than birds (Avise and Aquadro 1982, Ward *et al.* 1992). Even sympatric species pairs that segregated 100% by male attraction to pheromones produced by virgin females in our studies (Powell and De Benedictis Part 2 this vol.) showed genetic distance coefficients of only 0.006-0.078 (Harvey 1994). We interpret this to mean that reproductive isolation among members of the group is only recently and not in all cases completely established.

On a phenogram using modified Roger's distance, *C. retiniana spaldingiana* grouped with the pine-feeders, *C. pinus* and *C. lambertiana ponderosana*, but *C. l. subretiniana* grouped with the Abietoideae-feeders (Harvey 1994). *C. murinana* (Hübner), a Palaearctic Abietoideae-feeder, grouped with non-conifer-feeding out-group *Choristoneura* species, suggesting that the conifer-feeders of the Palaearctic and Nearctic had separate origins. This is corroborated by the markedly different pheromone chemistry of the European species (Priesner *et al.* 1980). This idea contradicts the hypothesis developed by Dang

(1992) based on male genitalia, in which the Palaearctic and Nearctic conifer-feeding species are considered species groups, linked by the form of the sacculus tip.

Sperling and Hickey (1994) provided strong evidence that *Choristoneura fumiferana* represents the oldest lineage within the species group, but they did not include any populations of *C. carnana, C. retiniana,* or *C. lambertiana*. Hence, the DNA evidence to date is not sufficiently comprehensive to determine relationships among the western species of the complex. It appears that all the western members are derived from a common ancestor, with its sister lineage represented by *C. fumiferana*. The DNA data suggest that both pine-feeding and secondary loss of aldehyde production in the pheromones developed independently in two or more species.

Both the DNA and allozyme evidence indicate geologically recent genetic divergence among the western species, which is reflected in the reproductive compatibility in confined interspecific pairs and by the species intermediacy situation in parts of the geographic range. Employing the sister group hypothesis posed by molecular data, we can speculate on speciational events in western members of the complex during the late Tertiary.

Sanders (1971) postulated that *C. retiniana (=viridis), C. orae* Freeman (an Abietoideae-feeding species in coastal British Columbia), and *C. pinus* Freeman (the eastern pine-feeding species) may have arisen from a common ancestor, because they share similar pheromones. The principal component was determined to be E-11-tetradecenyl acetate (Sanders *et al.* 1974), in contrast to E-11-tetradecenal shared by the other Abietoideae-feeding species. Later studies showed that different blends of E- to Z-11-tetradecenyl acetates induce male response in *C. retiniana* (Daterman *et al.* 1984) and *C. lambertiana* (Daterman *et al.* 1983, summarized in Carolin *et al.* 1987; Daterman *et al.* Part 3 this vol.).

Subsequently, Silk and Kuenen (1986, 1988) demonstrated that E/Z-11-14:acetates are the major components in all coniferophagous *Choristoneura* for which female sex pheromone glands have been chemically analyzed. Biochemical evidence indicates that the species producing aldehydes (*C. fumiferana, C. biennis, C. occidentalis,* and *C. carnana*) store gland-synthesized acetates, which are the direct biosynthetic precursors to the emitted aldehyde pheromones (Morse and Meighen 1984, 1986). This sequence, together with the vastly more comprehensive occurrence of acetate-based pheromones in all tribes of Tortricinae, as well as in Olethreutinae (Roelofs and Brown 1982), suggests that the aldehyde-derivative mode of pheromone production is the apomorphic state in Tortricidae generally and in the conifer-feeding *Choristoneura* specifically. Presumably it evolved independently in tthe aspen-feeding *C. conflictana* (Walker), in *Argyrotaenia citrana* (Fernald) (Archipini), and in *Acleris* and *Croesia* (Tortricini).

Therefore, the shared, acetate-based pheromone is interpreted to be a plesiomorphic resemblance among *C. pinus, C. orae,* and *C. retiniana* and of no value in reconstructing evolutionary branching patterns. Nevertheless, following the line of speculation introduced by Sanders (1971), we can consider a split into acetate- and aldehyde-producing groups as the fundamental speciational event (Fig. 7.3). However, this requires setting aside the molecular divergence data as an indicator of relationships and implies independent loss of the aedeagal spiculae (Dang 1992) at least twice—three times if *C. orae* is assumed to be a sister of *C. pinus* rather than of *C. lambertiana.*

Conversely, the dichotomy between Pinoideae-feeding and Abietoideae-feeding might reflect a fundamental mode for speciation in western *Choristoneura.* A speciation event involving adaptation to *Pinus* may have occurred in both an eastern lineage that retained the plesiomorphic aedeagal spicules (*C. pinus*) and a western North American lineage (*C. lambertiana*) (Fig. 7.4). The latter event might have taken place in the Rocky Mountains as early as the Miocene, with Pacific Coast populations of Abietoideae- and *Pinus*-feeding lineages evolving much later. Alternatively, a host plant switch might have occurred only once, which would have required not only independent loss of the aedeagal spicules in *C. lambertiana* and the Abietoideae-feeding lineage, but parallel narrowing of the forewing and correlated morphological changes (Fig. 7.5). Both hostplant-based scenarios imply homoplasy in parallel evolution of aldehyde-producing pheromones, in eastern and western complexes of *Choristoneura.*

Choristoneura retiniana may have originated as a derivative of *C. lambertiana*, or the reverse, by host race formation in sympatry or in peripheral populations when the arid White Fir forest of the Great Basin diverged during the Pliocene, which is equally consistent with pheromone or host tree dichotomy as a basis for sister groups. Such a scenario implies independent host transfer to *Pinus* by ancestors of *C. pinus* and *C. lambertiana*, and a derived adaptation to *Abies concolor* by *C. retiniana.* Morphological similarities between *C. retiniana* and *C. lambertiana*—narrow forewing, tendencies to fewer hindleg basitarsal spines and broader uncus tip, relative to *C. carnana-occidentalis* populations (De Benedictis Part 4 this vol.)—are corroborated by this theory.

Newly molted pupae of conifer-feeding *Choristoneura*, prior to darkening of the cuticle, show a sex-controlled polymorphism in haemolymph color, yellow or green, which is the result of absence or presence of blue bile pigments (Stehr 1959, 1964). The general occurrence of these pigments in other Lepidoptera and Orthoptera, and its variability in response to ecological or physiological conditions, led Stehr (1959) to speculate that the pigments reflect some role in developmental or physiological processes. The patterns of this polymorphism in *Choristoneura* are difficult to assess in a phylogenetic context, however, because out-group *Choristoneura* and other Tortricinae have not been studied, the polymorphism is geographically variable within species, and we lack data for *C. carnana* and *C. retiniana spaldingiana.*

The haemolymph is green in both sexes of *C. retiniana* (Warner Mountains population). It is green in females but yellow in males of *C. pinus pinus* (Ontario, Manitoba, Saskatchewan), while all other conifer-feeding *Choristoneura* examined possess yellow haemolymph in a proportion of both sexes, with a higher proportion in males (Harvey and Stehr 1967). *C. fumiferana, C. biennis, C. occidentalis,* and *C. orae* have 33-68% yellow in females, 53-90% in males; while *C. lambertiana* (Colorado and Wyoming populations) have 75% yellow females, 84% yellow males, but *C. l. subretiniana* (Lake Tahoe population) and *C. pinus maritima* have 30% and 57% yellow females and 100% yellow males. The Palaearctic *C. murinana* has the highest proportion of yellow in females among the *Choristoneura* studied, 80%, and all yellow males.

If we assume that yellow haemolymph is ancestral, as did Stehr (1959), which is a state unknown in any *Choristoneura, C. murinana* has diverged least; but from allozyme data and pheromone chemistry, cited above, we assume that haemolymph polymorphism has diverged independently in Nearctic *Choristoneura*. Males of *C. pinus* and *C. l. subretiniana* also retain the plesiomorphic state, while females of all species are either polymorphic or entirely green, and the most derived state is an autapomorphy in *C. retiniana*. If green haemolymph was ancestral, *C. retiniana* retains the plesiomorphic state, while all others have diverged to varying degrees. If we ignore *C. carnana* or assume it is similar to *C. occidentalis* in pupal haemolymph polymorphism, because it is in other characteristics, the distribution of haemolymph polymorphism suggests that *C. pinus* and at least one population of *C. lambertiana* have retained the ancestral type in males and that *C. retiniana* is the most derived species, which is in keeping with the scenario deduced from paleobotanic data and other lines of evidence, outlined above.

Neither of the two alternative pathways to speciation proposed above as a basis to determine sister groups in the complex, acetate vs. aldehyde (Fig. 7.3) or Pinoideae vs. Abietoideae (Figs. 7.4, 7.5), is affected by inclusion of haemolymph polymorphism data. Polymorphism in males occurs in *C. fumiferana*, the earliest lineage according to molecular data (Sperling and Hickey 1994), as well as in Abietoideae- and Pinoideae-feeders and acetate and aldehyde producers. Therefore the available evidence suggests a consensus phylogeny portrayed in Figs. 7.4 and 7.6. *C. fumiferana* is perceived as the sister to the rest of the North American conifer-feeding complex, with divergence of *C. pinus* from the western clade, and divergence of *C. retiniana* + *lambertiana* as early as the Miocene. Differentiation in the *C. orae-carnana-occidentalis* complex presumably took place in the late Tertiary and Quaternary.

ORIGIN OF THE SPECIES INTERMEDIACY COMPLEX IN OREGON

We postulate that the heterogeneous populations of *Choristoneura* in interior northern California and Oregon are the result of recent (ca. 7,000 yr) recontact of Rocky Mountain and coastal populations of *Pseudotsuga* and its *Choristoneura* species from the two regions, together with invasion of that area by *Abies concolor* and *C. retiniana* from more southerly refugia, during post-Wisconsin time.

Hybridization of the populations corresponding to stabilized, sympatric species in the south (*C. carnana* and *C. retiniana*) may have been enhanced during massive population outbreaks that occurred when Douglas-fir and White Fir forests reinvaded periglacial areas during the spread and recontact of coastal and interior races of Douglas-fir. In conditions of crowded individuals, of one or both species, confusion of pheromone signals and accidental matings with the other species would be likely, based upon what we know of interspecific matings of *Choristoneura* in laboratory situations (Sanders *et al.* 1977, Liebhold *et al.* 1984, Daterman *et al.* Part 3 this vol., Powell and De Benedictis Part 2 this vol.).

We consider the opposite hypothesis, hybridization beginning as a result of rarity, to be less tenable. In this scenario, if a few individuals of one or both species initially colonized the new habitat, there would be selection against males that could not locate females. Therefore males might respond to the pheromone of females of another species in the vicinity. Our observations of *Choristoneura* in non-hybridizing areas of California, Nevada, and Utah, usually in sparse population levels (Powell and De Benedictis Part 2 this vol.), demonstrated that 100% of the males selected the "right" pheromone, even when one species was very rare compared to the sympatric congener. If both sexes of one or both species were rare, few matings might occur, although we do not know the maximum distance from which males are attracted. If one or both species were extremely abundant, occasional "wrong" matings might result, as they do in confined interspecific pairs. We assume this is the result of crowded conditions, not an ability of males to change their response. Once hybrid offspring were present, backcrosses were likely, and would have continued, even in periods of low population numbers, or when one species is vastly outnumbered.

It is not clear whether individual hybrid females possess a broader chemical range in pheromones produced or whether individual males respond to only a narrow range of the parental pheromones. Sanders *et al.* (1977) found that F_1 males respond in equal numbers to females of both parent species. Studies by Liebhold *et al.* (1984) and De Benedictis *et al.* (Part 5 this vol.) suggest dominance in inheritance of *C. occidentalis* pheromone over that of *C. retiniana*. After multiple backcrosses, however, matings may be essentially at random.

TAXONOMY

The following diagnoses summarize characters measured by De Benedictis (Part 4 this vol.) (males only) and color patterns of both sexes observed in other material examined, which was in excess of 6,000 specimens (ca 68% UCB, 11% USNM, 11% LACM). Specimens were examined from the following collections: American Museum of Natural History, New York (AMNH); British Museum (Natural History), London (BMNH); California Academy of Sciences, San Francisco (CAS); California State Department of Food and Agriculture, Sacramento (CDFA); Canadian National Collection, Ottawa (CNC); Colorado State University, Fort Collins (CSU); Los Angeles County Museum of Natural History (LACM); San Diego Natural History Museum (SDNH); U.S. Forest Service, Pacific Northwest Experiment Station, Corvallis, Ore. (USFC); U.S. Forest Service, Pacific Southwest Experiment Station, Berkeley (USFB); U.S. National Museum of Natural History, Washington, D.C. (USNM); Essig Museum of Entomology, University of California, Berkeley (UCB); University of California, Davis (UCD); University of Idaho, Moscow (UI); Washington State University, Pullman (WSU).

Abbreviations: FW = Forewing; HW = Hindwing; n = number of exemplars measured or units counted (e. g., spines).

Choristoneura retiniana (Walsingham)
(Fig. 7.8)

Lozotaenia retiniana Walsingham, 1879, Illus. Lepid. Het. Brit. Mus. 4:12.
Choristoneura retiniana; Freeman, 1958, Canad. Entomol. Suppl. 7:36.
Choristoneura lambertiana lindseyana Obraztsov, 1962, Amer. Mus. Novitates 2101:16.
Choristoneura lindseyana; Powell, 1964, Univ. Calif. Publ. Entomol. 32:174 (synonymy).
Choristoneura viridis Freeman, 1967, Canad. Entomol. 99:452; Powell, 1980, USDA For. Serv. Gen. Tech. Rept. PNW-100:3 (synonymy).

Type data: *retiniana*: Mt. Shasta, California, August 1871 (BMNH). *lindseyana*: Davis Creek, 3 mi. E, Modoc Co., Calif., July 15-23, 1922 (NMNH). *viridis*: Bidwell Creek near Ft. Bidwell, Modoc Co., Calif., July 27, 1961 (CNC).

Diagnosis: FW length in male 9.8-16.6 mm ($\bar{x}$ =12.2 mm in southern populations, to 14.4 mm in Oregon); breadth 2.23-2.71 ($\bar{x}$ = 2.52) in southern populations to moderately broad ($\bar{x}$ = 2.41) in Oregon. Adults dimorphic in both sexes, FW either pale tan, without reticulation, or rust-brown to orange reticulated with dark lines. Northern California-Oregon populations (*C. retiniana retiniana*) more variable in color and FW pattern than in the Sierra Nevada and Great Basin Ranges (*C. r. spaldingiana*), with a decreasing proportion of the tan morph northward and increasing sexual dimorphism, the reticulated morph males mostly gray-brown to reddish brown, females reddish brown to orange. Probably these populations represent the southern limit of a broad hybrid zone with *C. occidentalis.*

Larva and pupa with weakly pigmented integument, appearing bright green when alive.

Sex pheromone: Principal component is E-11-tetradecenyl acetate; at 100% this compound is not attractive to males in California but attracts some males in southern Oregon. A blend of 90% 92:8 E:Z-tetradecenyl acetate and 10% 90:10 E:Z-tetradecenyl alcohol was attractive to males of all populations tested.

Geographical distribution: Mountains of southern Oregon east of the Cascade Range, California, Nevada, and Utah (Fig. 7.8).

Larval food plants: Abies concolor and *A. magnifica.*

Choristoneura retiniana retiniana (Walsingham), new status
(Figs. 7.8, 7.11 - 7.28)

Diagnosis: FW length in male 10.9-16.6 mm $\bar{x}$ =13.3 mm in Calif. to 14.4 mm in Ore.); breadth 2.23-2.71 ($\bar{x}$= 2.49 in Calif. to 2.41 in Ore.). Color and FW pattern extremely variable, becoming indistinguishable from *C. occidentalis* in the hybrid zone in Oregon. Unmottled, tan morph uncommon (20% or less of the population) in California to absent in Oregon. FW dark mottling ranging from pale ochreous-orange to dark gray-brown, with males approx. 80% reddish- to gray-brown, females nearly all orange to reddish-brown (Modoc populations). HW pale gray to dark gray-brown, occasionally tan at apex. Hindleg basitarsi moderately spinose, 9-26 ($\bar{x}$ =14.1 in Calif. to 17.9 in Ore.). Uncus shape similar to *C. occidentalis*, tip slightly broader than basal width, constriction ca. 0.7 basal width.

Geographical distribution: Klamath Ranges, Mt. Shasta area, Siskiyou Co., and Manzanita and Warner Mountains, Modoc Co., California, and adjacent areas of southern Oregon (Fig. 7.8). The northern limit of *C. retiniana* roughly corresponds to the line of hybridization between *Abies grandis* and *A. concolor* at the northern limit of the latter (Hamrick and Libby, 1972).

Choristoneura retiniana spaldingiana Obraztsov, new status
(Figs. 7.8, 7.29 - 7.32)

Choristoneura spaldingiana Obraztsov, 1962, Amer. Mus. Novitates 2101:6.

Type data: Provo, Utah, July 30, 1909 (AMNH).

Diagnosis: FW length in male 9.8-13.8 mm ($\bar{x}$ =12.2 mm); breadth 2.33-2.72 ($\bar{x}$=2.52). FW color dimorphic: 40-100% of each population tan (consistent from year to year but geographically variable, see De Benedictis Part 4 this vol.), sometimes with faintly defined, longitudinal, brown streaks but without dark reticulation or transverse mottling; remainder orange or orange-brown with a reticulate pattern of reddish brown or dark brown lines or strigulae, sometimes in part reduced so as to suggest pale bands through the reticulate pattern; no sexual dimorphism in color or frequencies of the two forms. HW pale gray, often tan at apex. Hindleg basitarsal spines significantly fewer than in *C. r. retiniana*, n = 6-15 ($\bar{x}$ =9.1), often fewer hindleg than midleg spines. Uncus weakly constricted, about 0.85 basal width, tip moderately broad, $\bar{x}$ =1.3 times basal width.

Geographic distribution: Higher elevations of North Coast Ranges, Sierra Nevada, Tehachapi, Transverse and Peninsular Ranges of California, Clark Mountains, Schell Creek and Snake ranges, Nevada, southern, central and Wasatch Mountains of Utah (Fig. 7.8).

Choristoneura carnana (Barnes and Busck)
(Fig. 7.9)

Tortrix carnana Barnes and Busck, 1920, Contr. Nat. Hist. Lepid. N. A. 4:214.

Choristoneura carnana; Freeman, 1958, Canad. Entomol. 90, Suppl. 7:36; Powell, 1980, USDA For. Serv. Gen. Tech. Rept. PNW-100:3.

Choristoneura lambertiana carnana; Powell, 1964, Univ. Calif. Publ. Entomol. 32:183.

Type data: Camp Baldy, San Bernardino Mts., California [=Camp Baldy (now Mt. Baldy P. O.), San Gabriel Mts., San Bernardino Co.] (NMNH).

Diagnosis: FW length in male 11.2-14.5 mm ($\bar{x}$ =13.1 mm); breadth 2.18-2.65 ($\bar{x}$ =2.40). FW color rust-reddish to orange-brown, with two broad, well defined, oblique, transverse bands of white or ochreous, sometimes fused along dorsal margin to termen. Populations in the North Coast Ranges (*C. carnana californica*) include individuals with the transverse bands less well defined and the pattern overlaid with some blackish reticulation, especially in females. Otherwise no sexual dimorphism in color and FW pattern. HW pale yellow-orange to orange except gray with orange apically in North Coast Ranges. Hindleg basitarsal spines moderately numerous, 10-24 ($\bar{x}$ =15.1). Uncus tip ca. 1.1 times, constriction about 0.75, the basal width, similar to *C. occidentalis*.

Larva and pupa with heavily pigmented integument, appearing brick red to dark reddish brown in life.

Sex pheromone: Not determined directly for typical *C. carnana*; identified in *C. carnana californica* by Daterman *et al.* (Part 3 this vol.). E-11-tetradecenal is the principal component; the synthetic lure, 100% blend of 92:8 E:Z-11-tetradecenal, developed for *C. fumiferana* and *C. occidentalis*, attracts *C. carnana* males of all populations tested.

Geographical distribution: Mountains of California at middle elevations, from below 500 m in the north to above 2,100 m in southern California (Fig. 7.9).

Choristoneura carnana carnana (Barnes and Busck)
(Figs. 7.9, 7.33 - 7.36)

Choristoneura carnana carnana; Powell, 1980, USDA For. Serv. Gen. Tech. Rept. PNW-100:3.

Diagnosis: FW length in male 11.4-13.4 mm ($\bar{x}$ =12.2 mm); breadth 2.32-2.59 ($\bar{x}$ = 2.44). FW in both sexes solid rust-red to orange-brown, with broad white or cream colored bands completely transecting the ground obliquely; bands fused on dorsal margin to termen, greatly reducing the ground color to a basal patch, a band distad of the median fascia, and a small triangular patch distad of subapical fascia. HW pale yellow-orange. Hindleg basitarsi less spinose (n = 10-15, $\bar{x}$ =12.4) than in other *C. carnana* populations and *C. occidentalis*.

Geographical distribution: Pine Mountain area in Ventura Co., San Gabriel Mountains of Los Angeles-San Bernardino Cos., and San Bernardino Mountains, San Bernardino Co., California (Fig. 7.9).

Larval foodplant: *Pseudotsuga marcrocarpa.*

Populations in the Sierra Nevada are similar to *C. carnana carnana* in phenotype and variation except the transverse bands of the FW are pale ochreous. The larvae feed on *P. menziesii* and occasionally on *Abies concolor* in mixed tree stands.

Similar moths with slightly longer FW and gray HW occur in the vicinity of Flagstaff, Arizona, and in the Sierra Madre Oriental of Coahuila, Mexico. These are characterized under the discussion of *C. occidentalis* (see below).

Choristoneura carnana californica Powell
(Figs. 7.9, 7.37 - 7.50)

Choristoneura lambertiana californica Powell, 1964, Univ. Calif. Publ. Entomol. 32:179.
Choristoneura carnana californica; Powell, 1980, USDA For. Serv. Gen. Tech. Rept. PNW-100:3.

Type data: Anderson Springs, Lake Co., California, June 24, 1958 (CAS).

Diagnosis: FW length in male 11.2-14.5 mm ($\bar{x}$ = 13.3 mm); breadth 2.18-2.65 ($\bar{x}$ = 2.39). Similar to *C. c. carnana*, FW transverse bands narrower, darker, often incomplete or broken; FW ground color, especially of females, tends to brownish, sometimes mottled and reticulated with darker scales. Males generally more russet-red, less mottled and lacking dark reticulation. HW usually gray, often pale yellow-orange distally; rarely more extensive yellow-orange. Hindleg basitarsal spines more numerous than in other populations of *C. carnana*, n = 11-24 ($\bar{x}$ =15.8).

Geographical distribution: California: North Coast Ranges from the Santa Cruz Mountains (San Mateo Co.) north to Del Norte County, and in the Trinity-Salmon Mountains and Mt. Shasta area, Siskiyou Co. (Fig. 7.9).

A less variable form, intermediate to *C. carnana carnana*, occurs from Plumas Co. southward along the west slope of the Sierra Nevada (Figs. 7.37 - 7.40). Northward from the type locality, sexual dimorphism in color and FW pattern gradually diminishes across the mountains north of the Central Valley until dimorphism is not evident in the Sierra Nevada, where the gray HW of the Coast Ranges is replaced by peach- or yellow-orange. Specimens from the northern part of the range tend to be darker with more broken FW banding. Size, basitarsal spination, and uncus shape remain more or less constant.

Larval foodplants: Pseudotsuga menziesii and in northern populations also *Abies concolor* in mixed tree stands.

Choristoneura occidentalis Freeman
(Figs. 7.9, 7.57 - 7.72)
Choristoneura occidentalis Freeman, 1967, Canad. Entomol. 99:451.

Type data: Yakima Indian Reserve, Monument Rd., Klickitat Co., Washington, July 7, 1961 (CNC).

Diagnosis (excludes populations A-C, described below): FW length in male 11.3-17.2 mm ($\bar{x}$ = 14.3 mm); breadth 2.12-2.60 ($\bar{x}$ =2.40). Diverse populations occur over a wide area of western North America, encompassing a range of phenotypes. Most populations variable, FW strongly mottled, ground color usually a mosaic of dark scales, brown to reddish brown or dark gray-brown. Males generally smaller and darker than females, with obvious sexual dimorphism in size and color in most populations. HW pale to dark gray. Hindleg basitarsal spines numerous, n = 10-29 ($\bar{x}$ =17.8). Uncus tip ca. 1.1 times, constriction, ca. 0.75 the basal width. Adults in northern populations tend to be larger, darker, and have more spinose basitarsi.

Mature larva in Rocky Mountains olive-brown to reddish brown dorsally with large, conspicuous, pale setal pinacula, pale orange to yellow subventrally and ventrally; head and thoracic shield tan to orange-brown, rarely dark brown (Harvey and Stehr 1967, Stevens *et al.* 1984). Pupa reddish brown. In the hybrid zone east of the Cascades in Oregon, larvae range from nearly unpigmented, appearing green when alive, to reddish brown with corresponding increase in pigmentation of head and thoracic shield and in the pupa.

Several populations in which males respond to OXY pheromone bait are more variable or exhibit phenotypes that are not typical of *Choristoneura occidentalis* in the Cascade Range and Rocky Mountains. The more divergent of these may be characterized as follows.

Population A (south-central to east-central Oregon) (Figs. 7.57 - 7.62): Males from the hybrid zone in Oregon vary phenotypically more than other populations in size and FW pattern but not in color; all are dark gray-brown (n=50+), with a few also tinged with reddish. Sexual dimorphism pronounced; females reared from brown or intermediate colored larvae are 70% bright rust to dark reddish brown.

Population B (Schell Creek and Snake Ranges, Nevada) (Figs. 7.69-7.70): Adults smaller (FW 11.2-14.8 mm; $\bar{x}$ =13.4 mm) with rather uniformly mottled or checkered FW, gray-brown in male, pale brown to orange-brown in female.

Population C (vicinity of Flagstaff, Coconino Co., Arizona): Adults small (FW 11.5-13.0 mm; ($\bar{x}$ =12.4 mm)) with FW distinctly banded, resembling *C. carnana*. At West Fork Oak Creek, SW of Flagstaff (elev. 2,000 m), the population sample (NMNH) also includes individuals that are more brown, with the FW ground color somewhat mottled, more extensive than in the *C. carnana*-like moths. A similar population occurs in Coahuila, Mexico, n = 3 specimens labeled "*Pseudotsuga macrolepis*" (UCB). At Hochderffer Hill, NW of Flagstaff (elev. 2,600 m), the same range of phenotypes occurs, along with moths that are more typical of *C. occidentalis* in FW pattern and color.

In the latter population and at the Jarbidge Wilderness area in northern Nevada, *C. occidentalis* are of two size classes, most males having FW length ca 13 mm, but others are 15+ mm.

Geographical distribution: Populations assigned to this name occur in Alberta, British Columbia, Washington, Oregon, and in the Rocky Mountains and Great Basin Ranges in Idaho, Montana, Nevada, northern Arizona, Wyoming, Colorado, and New Mexico (Fig. 7.9). Stehr (1967) illustrated a collection point for *C. occidentalis* in the northern Sierra Nevada, and Freeman (1967) listed California, presumably based on the same record, which must have been misidentified *C. retiniana* or *C. carnana.*

Sex pheromone: E-11-tetradecenal is the principal component. A synthetic lure consisting of 100% blend of 92:8 E:Z-11-tetradecenal is effective, while E-11-tetradecenyl acetate baits are not, in all populations tested in Oregon west of the Cascade Range, Great Basin Ranges of Nevada, Utah and northern Arizona, and presumably the same response occurs in the Rocky Mountains. In hybridizing populations east of the Cascade Range crest in south and central Oregon, males respond to either E- and Z-11-14 Ac or E- and Z-11-14 Al as the principal component.

Larval foodplants: *Pseudotsuga menziesii* is reported to be the major host, but other Pinaceae, including *Tsuga heterophylla*, *Larix occidentalis,* and various species of *Abies*, *Picea*, and *Pinus* are used (Stehr 1967), especially in high density populations.

Choristoneura lambertiana (Busck)
(Fig. 7.10)

Tortrix lambertiana Busck, 1915, Proc. Entomol. Soc. Wash. 17:86.
Choristoneura lambertiana; Freeman, 1958, Canad. Entomol. 90, Suppl. 7:36.

Type data: "Oakland, Oreg." reared from *Pinus lambertiana*, June 7, 1914 (NMNH) [label data from the type specimens and associated Hopkins U.S. data indicate that the collection was made at Ashland, Jackson Co., Oregon, June 18, 1914 (Powell 1964)].

Diagnosis: Adults generally small (FW in male 10.6-14.6 mm; $\bar{x}$ =12.7 mm); FW breadth (2.26-2.77; $\bar{x}$ =2.51). FW pattern geographically variable but relatively monomorphic within populations, without appreciable sexual dimorphism; brick red- or rust-orange with well developed pattern of irregular transverse bands of gray or whitish, to pale ochreous, with washed out pattern, no transverse bands. HW whitish to pale gray.

Hindleg basitarsal spines geographically variable, moderately sparse to moderately numerous (n=10-21). Uncus tip moderately broad to broad, 1.0-1.9 times basal width; constriction moderate to weak, 0.60-0.96 basal width.

Larva (Sierra Nevada *Choristoneura l. subretiniana* and Charleston Mtn. *C. l. ponderosana* in UCB): dark reddish brown to pale rust dorsally with conspicuous pale pinacula; head reddish tan to reddish amber, thoracic shield in male pale to dark reddish brown, in female, nearly all tan to pale reddish tan (Harvey and Stehr 1967). Pupa reddish brown.

Geographical distribution: Widespread in pine regions of western North America, at moderately high elevations in California, Oregon, Idaho, Montana, Wyoming, western South Dakota, Colorado, Utah, northern Arizona, and New Mexico (Fig. 7.10). We have not examined specimens from eastern Washington and Canada, but Freeman (1967, Fig. 16), Stehr (1967 Fig. 1), and Harvey and Stehr (1967) reported *C. lambertiana* from southern Alberta; and Harvey (1985: Fig. 3) included Washington and British Columbia within the range of this species.

A population near Vancouver, British Columbia, studied by Gray and Slessor (1989) in associartion with Scots Pine plantations, may be the same as the "Pacific Northwest Coastal Populations" in Oregon and Washington recorded here and by Harvey (1985) and De Benedictis (Part 4 this vol.). Gray and Slessor found the pheromone chemistry to be unique but attractive to males of *C. pinus*, which suggests a relationship to that species rather than to *C. lambertiana.*

Sex pheromone: E-11-tetradecenyl acetate is the principal component. A synthetic blend of 90% 60:40 E:Z-11-tetradecenyl acetate and 10% 60:40 or 67:33 E:Z-11-tetradecenyl alcohol attracted males in geographically disjunct and phenotypically differing populations.

Larval foodplants: Primarily *Pinus*; larvae occasionally occur on Abietoideae.

Choristoneura lambertiana lambertiana (Busck)
(Figs. 7.10, 7.73)

Choristoneura lambertiana lambertiana; Obraztsov, 1962, Amer. Mus. Novitates 2101:11.

Diagnosis: Few specimens exist. FW length in males 10.6-13.2 mm ($\bar{x}$ = 11.8 mm, 4 n); breadth 2.25-2.50 ($\bar{x}$ = 2.36). FW pale orange with narrow, usually incomplete pale ochreous or whitish bands. HW pearly white tinged with yellow-orange or gray, especially apically. Basitarsal spine counts and uncus shape resemble those of *C. lambertiana subretiniana.*

Geographical distribution: In the restricted sense interpreted here, the typical race is known only from the type locality and the adjacent Siskiyou Mountains in California (Fig. 7.10). The concept of this subspecies has been variously enlarged to include populations of the northern Rocky Mountains in Idaho, Montana, and Wyoming (Obraztsov 1962, Stehr 1967, Stevens *et al.* 1977). These show phenotypic, host plant, and geographic affinities to *C. l. subretiniana*, rather than to *C. l. lambertiana* in the strict sense.

Larval foodplant: Pinus lambertiana.

Choristoneura lambertiana subretiniana Obraztsov
(Figs. 7.10, 7.75 - 7.78)

Choristoneura subretiniana Obraztsov, 1962, Amer. Mus. Novitates 2101:9.
Choristoneura lambertiana subretiniana; Powell, 1964, Univ. Calif. Publ. Entomol. 32:181.

Type data: Monachee [Meadows], Tulare Co., California, 8,000 ft., July 16-23 [year unknown] (NMNH).

Diagnosis (includes northern Rocky Mountains populations): FW length of male 10.6-13.9 mm ($\bar{x}$ =12.2 mm); breadth 2.26-2.77 ($\bar{x}$ =2.46). Northern populations average slightly smaller, with broader FW. FW ground color somewhat mottled, brick red in California and Oregon, becoming orange to pale yellow-brown northward; transverse bands irregular but conspicuous, shining gray in the Sierra Nevada to whitish in the northern Rocky Mountains. Basitarsal spines average about 14 (n = 10-17) on both the hind and middle legs in all populations. Uncus tip moderately broad, about 1.3 times basal width (range 1.0-1.6); constriction width averages about 0.75 basal width (range 0.62-0.96).

Geographic distribution: Higher elevations and east slope of the Sierra Nevada, Warner Mountains in Modoc Co., California, and eastern Oregon. Atypical populations with the FW banding whitish occur in Idaho and western Montana (Fig. 7.10) and into eastern Washington and southern British Columbia (Harvey 1985: Fig. 3).

Larval foodplants: *Pinus contorta*, *P. ponderosa*; rarely *P. albicaulis*, *Abies concolor* and *A. magnifica* in mixed stands with lodgepole pine.

Choristoneura lambertiana ponderosana Obraztsov
(Figs. 7.10, 7.79 - 7.82)

Choristoneura lambertiana ponderosana Obraztsov, 1962, Amer. Mus. Novitates 2101:14.

Type data: Sugar Loaf, [Boulder Co.] Colorado, July 8, 1937 (NMNH).

Diagnosis: FW length in male 11.5-14.6 mm ($\bar{x}$ =13.2 mm); breadth 2.39-2.77 ($\bar{x}$ = 2.56). FW color pattern pale brownish orange, reduced to scattered blotches or mottling on a pale tan ground; underlying scales with a silvery gray sheen. HW pale gray or gray-brown. Basitarsal spines average 13.3 (n = 10-18), slightly fewer than in *C. l. subretiniana*. Uncus tip often greatly widened, averaging 1.6+ times basal width (range 1.26-1.92); constriction width about 0.8 times basal width (range 0.60-0.96).

Moths of the *C. l. ponderosana* phenotype and those of northern Rocky Mountain *subretiniana* fly together with intermediates in western and south-central Wyoming.

Geographical distribution: Southern Montana, Wyoming, western South Dakota, east slope of the Rocky Mountains in Colorado. Populations in southern Utah, northern Arizona, and Clark Mountains in southern Nevada are included, although the forewing phenotype shows indications of similarity to *C. l. subretiniana* of the northern Rocky Mountains (Fig. 7.10).

Larval foodplants: Pinus ponderosa; also recorded on *P. flexilis* in Wyoming (Stevens *et al.* 1977).

Choristoneura lambertiana (or *pinus*)
Pacific Northwest Coastal Populations
(Figs. 7.10, 7.74)

Diagnosis: FW length in male 11.5-13.9 mm ($\bar{x}$ = 12.7 mm); breadth 2.30-2.65 ($\bar{x}$ = 2.48). FW dark gray-brown densely mottled with reddish brown; transverse bands only weakly indicated. HW dark gray-brown. Basitarsi more spinose (n = 12-21, $\bar{x}$ = 16.2) than in other *C. lambertiana* populations. Uncus tip ca 1.1 times wider than basal width (range 0.79-1.27); constriction about 0.75 basal width (range 0.59-0.91).

Geographical distribution: Coastal Washington and Oregon (Fig. 7.10).
Larval foodplant: Pinus contorta.

ACKNOWLEDGMENTS

We thank the people who were instrumental in development of this study, acknowledged at the beginning of this volume. The evolutionary scenario was in part developed through discussions with W. J. A. Volney, who reviewed a draft of that part of our ms in 1984. Cooperation by curators of the collections listed above enabled use of specimens in their care, as named in Part 4 (De Benedictis 1955). Funding from the CANUSA-West Program supported our research during parts of 1978-79 and 1980-81, as did Grant No. BSR-8411459 from the National Science Foundation during 1985-1988.

LITERATURE CITED

Avise, J. C., and C. F. Aquadro. 1982. A comparative summary of genetic distances in the vertebrates. Evol. Biol. 15:151-185.

Axelrod, D.I. 1944. The Sonoma flora (California) and the Pliocene sequence in central California. Carnegie Inst. Wash. Publ. 553:167-224.

Axelrod, D.I. 1966. The Pleistocene Soboba flora of southern California. Univ. Calif. Publ. Geol. Sci. 60:1-79.

Axelrod, D.I. 1976. History of the coniferous forests, California and Nevada. Univ. Calif. Publ. Botany 70; 62 pp.

Axelrod, D.I., and W.S. Ting. 1960. Late Pliocene floras east of the Sierra Nevada. Univ. Calif. Publ. Geol. Sci. 39:1-118.

Axelrod, D.I., and W.S. Ting. 1961. Early Pleistocene floras from the Chagoopa Surface, southern Sierra Nevada. Univ. Calif. Publ. Geol. Sci. 39:119-194.

Ayala, F.J., M.L. Tracey, D. Hedgecock, and R. C. Richmond. 1974. Genetic differentiation during the speciation process in *Drosophila*. Evolution 28:576-592.

Baker, R.G. 1983. Holocene vegetational history of the western United States. *In:* H.E. Wright (ed.), Late Quaternary Environments of the United States, vol. 2:109-127. Univ. Minnesota Press

Carolin, V.M., J.A. Powell, G.E. Daterman, and M.W. Stock. 1987. Budworm taxonomy. *In*: M.H. Brookes *et al.* (tech. coord.), Western Spruce Budworm. USDA Tech. Bull. 1694:18-24.

Chaney, R.W. (ed.). 1944. Pliocene floras of California and Oregon. Carnegie Inst. Wash. Publ. 553; 407 pp.

Cole, K.L. 1982. Late Quaternary zonation of vegetation in the eastern Grand Canyon. Science 217:1142-1145.

Condit, C. 1944. The Remington Hill flora and the Table Mountain flora. Carnegie Inst. Wash. Publ. 553:21-90.

Crandell, D.R. 1965. The glacial history of western Washington and Oregon. *In*: H.E. Wright and D.G. Frey (eds.), The Quaternary of the United States. Princeton University Press; x + 922 pp.

Dang, P.T. 1985. Key to adult males of conifer-feeding species of *Choristoneura* Lederer (Lepidoptera: Tortricidae) in Canada and Alaska. Canad. Entomol. 117:1-5.

Dang, P. T. 1992. Morphological study of male genitalia with phylogenetic inference of *Choristoneura* Lederer (Lepidoptera: Tortricidae). Canad. Entomol. 124:7-48.

Daterman, G.E., H.T. Cory, G.D. Davis, Jr., R.E. Stevens, and D.G. Fellin. 1983 (unpubl.). A preliminary report on the sex pheromones of three pine-feeding budworms of the *Choristoneura lambertiana* Busck complex in the western United States. Forestry Sci. Lab., Corvallis, Ore.; 17 pp.

Daterman, G.E., H.T. Cory, L.L. Sower, and G.D. Davis, Jr. 1984. Sex pheromone of a conifer-feeding budworm, *Choristoneura retiniana* Walsingham. J. Chem. Ecol. 10:153-160.

Daterman, G.E., L.L. Sower, R.E. Stevens, and D.G. Fellin. 1995. Pheromone chemistry and response behavior of conifer-feeding *Choristoneura* in the western United States. Part 3 this volume.

De Benedictis, J.A. 1995. Phenetic studies of spruce budworm populations (*Choristoneura* species). Part 4 this volume.

De Benedictis, J.A., A.M. Liebhold, and J.A. Powell. 1995. Studies of pheromone attraction and inheritance, using virgin females of *Choristoneura* species in the central Sierra Nevada, California. Part 5 this volume.

Denton, G.H., and T.J. Hughes (eds.). 1981. The last great ice sheets. Wiley, New York; xviii + 484 pp. (+ maps, foldouts, bd. in 2nd vol.).

Flint, R.F. 1971. Glacial and Quaternary geology. Wiley, New York; xii + 892 pp.

Freeman, T.N. 1967. On coniferophagous species of *Choristoneura* (Lepidoptera: Tortricidae) in North America. I. Some new forms of *Choristoneura* allied to *C. fumiferana.* Canad. Entomol. 99:449-455.

Gray, T. G., and K. N. Slessor. 1989. Morphology, life history and identification of sex pheromone components of an undescribed species of *Choristoneura* (Lepidoptera: Tortricidae) on Scots Pine in British Columbia. J. Entomol. Soc. Brit. Columbia 86:39-47.

Hamrick, J.L., and W.J. Libby. 1972. Variation and selection in western U.S. montane species. I. White Fir. Silvae Genet. 20:29-35.

Harvey, G.T. 1985. The taxonomy of the coniferophagous *Choristoneura* (Lepidoptera: Tortricidae). *In*: C.J. Sanders *et al.* (eds.), Recent advances in spruce budworms research, pp. 16-48. Canadian Forest Service, Ottawa.

Harvey, G. T. 1994. Genetic relationships among *Choristoneura* species (Lepidoptera: Tortricidae) in North America. Canad. Entomol. (in press).

Harvey, G.T., and G. Stehr. 1967. On coniferophagous species of *Choristoneura* (Lepidoptera: Tortricidae) in North America. 3. Some characters of immature forms helpful in the identification of species. Canad. Entomol. 99:464-481.

Liebhold, A.M., W.J.A. Volney, and W.E. Waters. 1984. Evaluation of cross-attraction between sympatric *Choristoneura occidentalis* and *C. retiniana* (Lepidoptera: Tortricidae) populations in south-central Oregon. Canad. Entomol. 116:827-840.

Little, E.L., Jr. 1953. Check list of native and naturalized trees of the United States (including Alaska). Agricultural Handbook 41, USDA, Washington, D. C.; 472 pp.

Martin, P.S., and P.J. Mehringer, Jr. 1965. Pleistocene pollen analysis and biogeography of the Southwest. *In:* H.E. Wright and D.G. Frey (eds.). The Quaternary of the United States, pp. 433-451. Princeton University Press.

Morse, D., and E. Meighen. 1984. Aldehyde pheromones in Lepidoptera: Evidence for an acetate ester precursor in *Choristoneura fumiferana.* Science 226:1434-1436.

Morse, D., and E. Meighen. 1986. Pheromone biosynthesis and role of functional groups in pheromone specificity. J. Chem. Ecol. 12:335-351.

Obraztsov, N.S. 1962. New species and subspecies of North American Archipini, with notes on other species (Lepidoptera: Tortricidae). Amer. Mus. Novitates 2101:1-26.

Péwé, T.L. 1983. The periglacial environment in North America during Wisconsin time. In: H.E. Wright Jr. (ed.), Late Quaternary environments of the United States. Vol. 1, S.C. Porter (ed.), The late Pleistocene, pp. 157-189. University of Minnesota Press.

Porter, S. C. (ed.) 1983. Vol. 1. *ibid. In:* H.E. Wright (ed.) *ibid.*, xiv + 407 pp.

Porter, S. C., K.L. Pierce, and T.D. Hamilton. 1983. Late Wisconsin mountain glaciation in the western United States. *In:* H.E. Wright, Jr. (ed.) *ibid.* Vol. 1. *ibid.*, pp.71-114.

Powell, J. 1964. Biological and taxonomic studies on tortricine moths, with reference to the species in California (Lepidoptera: Tortricidae). Univ. Calif. Publ. Entomol. 32:1-317.

Powell, J.A. 1980. Evolution of larval food preferences in microlepidoptera. Ann. Rev. Entomol. 25: 133-159.

Powell, J. A., and De Benedictis. 1995. Biological relationships: Host tree preferences and isolation by pheromones among allopatric and sympatric populations of western *Choristoneura.* Part 2 this volume.

Priesner, E., H. Bogenshutz, and H. Arn. 1980. A sex attractant for the European fir bud moth, *Choristoneura murinana.* Z. Naturforsch. 35: 390-398.

Roelofs, W.L., and R.L. Brown. 1982. Pheromones and evolutionary relationships of Tortricidae. Ann. Rev. Syst. and Ecol. 13:395-422.

Sanders, C. J. 1971. Sex pheromone specificity and taxonomy of budworm moths (*Choristoneura*). Science 171:911-913.

Sanders, C. J., G.E. Daterman, and T.J. Ennis. 1977. Sex pheromone responses of *Choristoneura* spp. and their hybrids (Lepidoptera: Tortricidae). Canad. Entomol. 109:1203-1220.

Sanders, C. J., G.E. Daterman, R. Shepherd and H. Cerezke. 1974. Sex attractants for two species of western spruce budworm, *Choristoneura biennis* and *C. viridis* (Lepidoptera: Tortricidae). Canad. Entomol. 106:157-159.

Sanders, C. J., R.W. Stark, E.J. Mullins, and J. Murphy (eds.). 1985. Recent advances in Spruce Budworms research. Proc. CANUSA Spruce Budworms Res. Symp., Bangor, Maine, 1984. Canadian Forest Service, Ottawa; 527 pp.

Silen, R.R. 1978. Genetics of Douglas-fir. USDA For. Serv. Res. Paper WO-35; 34 pp.

Silk, P.J., and L.P.S. Kuenen. 1986. Spruce budworm (*Choristoneura fumiferana*) pheromone chemistry and behavioral responses to pheromone components and analogs. J. Chem. Ecol. 12:367-383.

Silk, P.J., and L.P.S. Kuenen. 1988. Sex pheromones and behavioral biology of the coniferophagous *Choristoneura*. Ann. Rev. Entomol. 33:83-101.

Sperling, F.A.H., and D.A. Hickey. 1994. Mitochondrial DNA sequence variation in the Spruce Budworm species complex (*Choristoneura*: Lepidoptera). Mol. Biol. Evol. 11: 656-665.

Stehr, G. 1959. Haemolymph polymorphism in a moth and the nature of sex-controlled inheritance. Evolution 13: 537-560.

Stehr, G. 1964. The determination of sex and polymorphism in microevolution. Canad. Entomol. 96: 418-428.

Stehr, G. 1967. On coniferophagous species of *Choristoneura* (Lepidoptera: Tortricidae) in North America. II. Geographic distribution in accordance with forest regions. Canad. Entomol. 99:456-463.

Stevens, R.E., T.K. Borg, and T.O. Thatcher. 1977. Notes on a pine-feeding budworm, *Choristoneura lambertiana ponderosana* (Lepidoptera: Tortricidae) in the Colorado Rockies. Canad. Entomol. 109:1269-1274.

Stevens, R.E., V.M. Carolin, and G.P. Markin. 1984. Lepidoptera associated with Western Spruce Budworm. USDA Forest Service Agric. Handbook no. 622.

Tsukada, M. 1982. *Pseudotsuga menziesii* (Mird.) Franco: Its pollen dispersal and late Quaternary history in the Pacific Northwest. Jap. J. Ecol. 32:159-187.

Volney, W.J.A. 1985. Comparative population biology of North American spruce budworms. *In*: C.J. Sanders *et al.* (eds.), Recent advances in spruce budworms research, pp. 71-84. Canadian Forest Service, Ottawa.

Volney, W.J.A., A.M. Liebhold, and W.E. Waters. 1984. Host associations, phenotypic variation, and mating compatibility of *Choristoneura occidentalis* and *C. retiniana* (Lepidoptera: Tortricidae) populations in south-central Oregon. Canad. Entomol. 116:813-826.

Wahrhaftig, C., and J.H. Birman. 1965. The Quaternary of the Pacific mountain system in California. *In*: H.E. Wright and D.G. Frey (eds.), The Quaternary of the United States, pp. 299-340. Princeton University Press.

Ward, R.D., D.O.F. Skibinski, and M. Woodwark. 1992. Protein heterozygosity, protein structure, and taxonomic differentiation. Evol. Biol. 26:73-159.

Wright, H.E., Jr. (ed.) 1983. Late Quaternary environments of the United States, vol. 2. The Holocene. University of Minnesota Press. xvii + 277 pp.

Table 7.1. Conifer-feeding Choristoneura in the Western United States

Taxon	Distribution
Choristoneura Lederer, 1859	
1. retiniana retiniana (Walsingham, 1879)	Siskiyou and Modoc Cos., Calif.
lindseyana Obraztsov, 1962	
viridis Freeman, 1967	
retiniana spaldingiana Obraztsov, 1962	Sierra Nevada, Tehachapi, So. Calif. Mts., Great Basin Ranges
2. carnana carnana (Barnes and Busck, 1920)	Transverse Ranges, So. Calif.
carnana Sierra Nevada populations	Westside Sierra Nevada below 1,600 m
carnana californica Powell, 1964	No. Coast Ranges to Mt. Shasta, Calif.
3. occidentalis Freeman, 1967	Brit. Columbia, Wash., western Ore., Idaho, Nev., Ariz., Mont., Wyo., Colo., N.Mex.
4. lambertiana lambertiana (Busck, 1915)	So. Ore., No. Calif.
lambertiana subretiniana Obraztsov, 1962	Sierra Nevada above 1,600 m, Warner Mts, E. Ore.
lambertiana N. Rocky Mts. populations	Idaho, Mont., Wyo.
lambertiana ponderosana Obraztsov, 1962	Wyo., W. S. Dak., E. side Rocky Mts., Colo., N.Mex., So. Utah, No. Ariz., So. Nev.
lambertiana [or pinus Freeman, 1953] coastal populations	Coastal Ore., Wash.

Table 7.2 Morphological, biochemical, and biological properties of conifer-feeding *Choristoneura* in the western United States.

	carnana	*occidentalis*	*retiniana*	*lambertiana*
Principal larval host	Pseudotsuga	Pseudotsuga and other Abietoideae	Abies concolor	Pinus
Pheromone principal component	aldehyde	aldehyde	acetate	acetate
Pupal haemolymph polymorphism	unknown	polymorphic both sexes	green both sexes	yellow male, polymorphic female in *subretiniana;* polymorphic both sexes in *ponderosana*
Mean forewing breadth	2.40	2.40	2.50 except Ore.	2.45-2.55
Mean no. basitarsal spines	12.4 in c. *carnana;* 16 in *californica*	18	14 (18 Ore) 9 in *spaldingiana*	13 in *ponderosana*; 14 in *subretiniana*
Uncus, tip: base ratio	1.1	1.1	1.1 in *retiniana;* 1.3 in *spaldingiana*	1.3 in *subretiniana*; 1.6 in *ponderosana*
Aedeagal spurs	absent	absent	absent	absent
Forewing polymorphism	none	red or brown in males	tan or reticulate both sexes	none

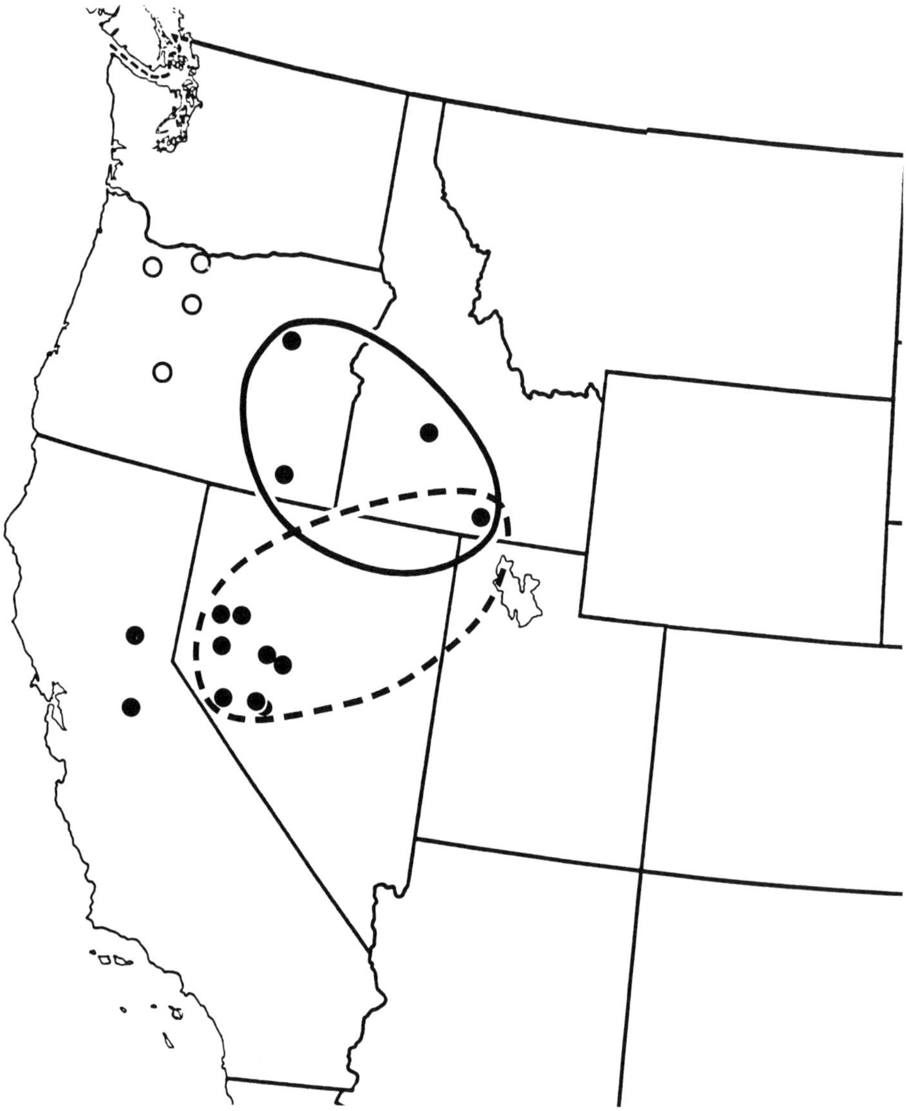

Fig. 7.1. Distribution of some fossil floras in the Great Basin region during early to mid Miocene (25 to 15 m yr BP). Open circles= coastal lowland floras with mixed deciduous hardwood forests; closed circles= floras containing montane conifer forests. Northern province (solid line) characterized by *Sequoia*, *Thuja*, and *Tsuga sonomensis* (= *heterophylla*); southern province (broken line) characterized by *Sequoiadendron*, *Tsuga mertensioides*, *Abies concoloroides*, and *Pinus florissanti* (=*ponderosa*) (modified from Axelrod 1976).

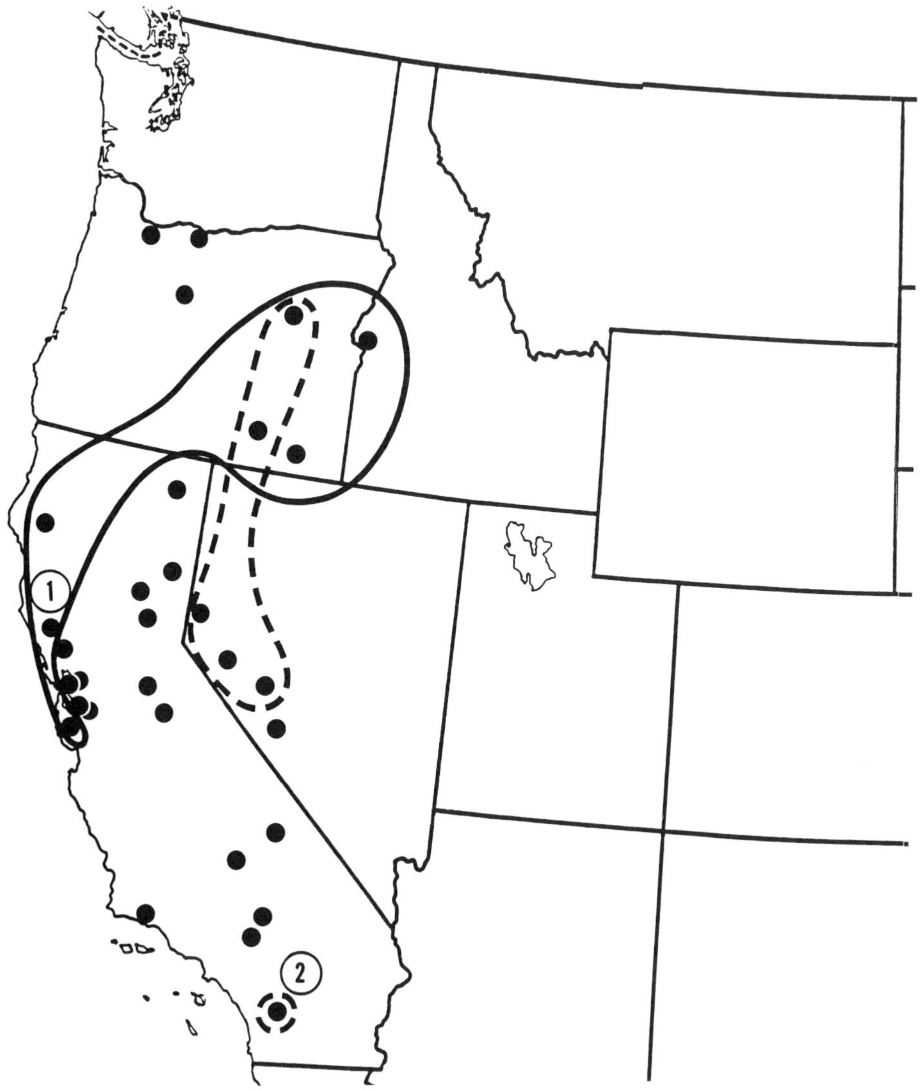

Fig. 7.2. Distribution of Mio-Pliocene and Pliocene floras in the Oregon, Nevada, and California region. Northern conifer forests (solid line) characterized by *Pseudotsuga sonomensis*, migrated coastward, while the interior forest, characterized by *Abies concoloroides*, spread southward into the drying interior. Symbols outside the curvilinear figures represent floras that had no Abietoideae. Only 14 of 31 Pliocene floras in the California-western Nevada region had any Pinaceae, including 4 with only *Pinus*. Sonoma Flora (1) included *Abies*, *P. sonomensis*, *Picea*, and *Tsuga*; Soboda Flora (2) had *A. concoloroides* and *P. premacrocarpa* at the end of the Pliocene (after Axelrod 1966, 1976, Chaney et al. 1944).

Figs. 7.3 - 7.5. Cladograms depicting hypothetical relationships and alternative pathways of speciational events in conifer-feeding *Choristoneura*.

Characters:	plesiomorphic	apomorphic
1. Larval hostplant	Abietoideae	Pinoideae
1a.		*Abies concolor*
2. Pheromone	acetate	aldehyde
3. Forewing breadth	broad	narrow
4. Aedeagal spicules	present	absent

Closed symbols indicate apomorph state; open symbols indicate reversal to plesiomorph state. Polarity of character transition is based on non-conifer-feeding *Choristoneura* species as out groups. Abietoideae is presumed ancestral because Pinoideae-feeding *Choristoneura* are restricted to regions that are broadly sympatric with Abietoideae. Decreased forewing breadth (length/width ratio) is correlated with tendencies to fewer basitarsal spines and broader uncus tip, the ranges of which overlap between species groups (De Benedictis Part 4 this vol.). These characters have not been measured for *C. orae*.

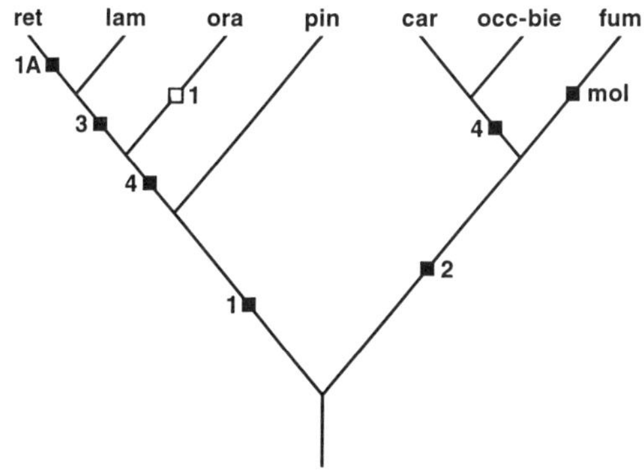

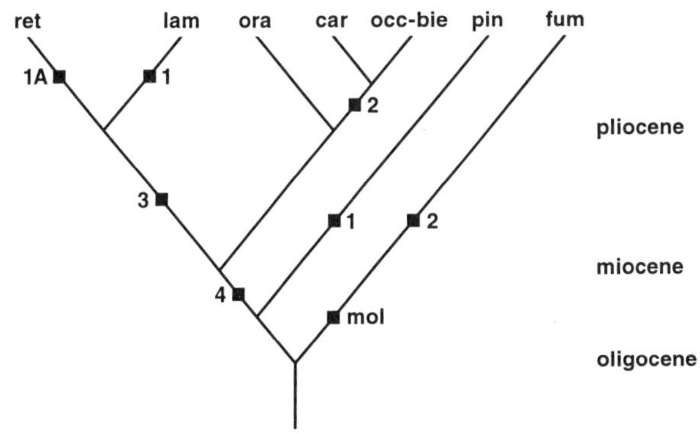

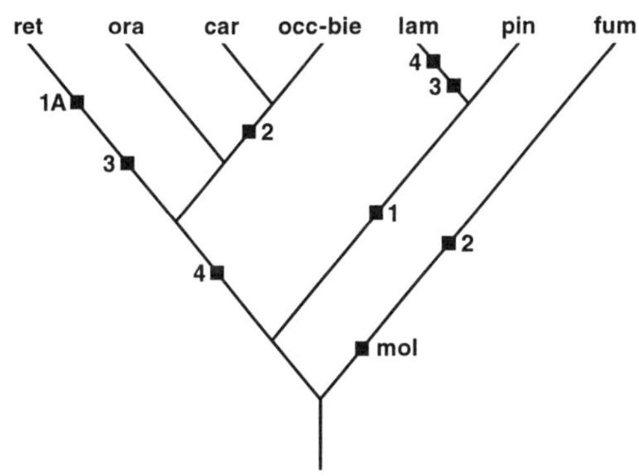

Fig. 7.3. Fundamental speciational event based on pheromone type.

Fig. 7.4. Fundamental speciational event based on molecular divergence (mol) (Harvey 1994, Sperling and Hickey 1994); assumes independent host switch to *Pinus* in eastern and western lineages.

Fig. 7.5. Fundamental speciational event based on molecular divergence, with a single host switch to *Pinus*.

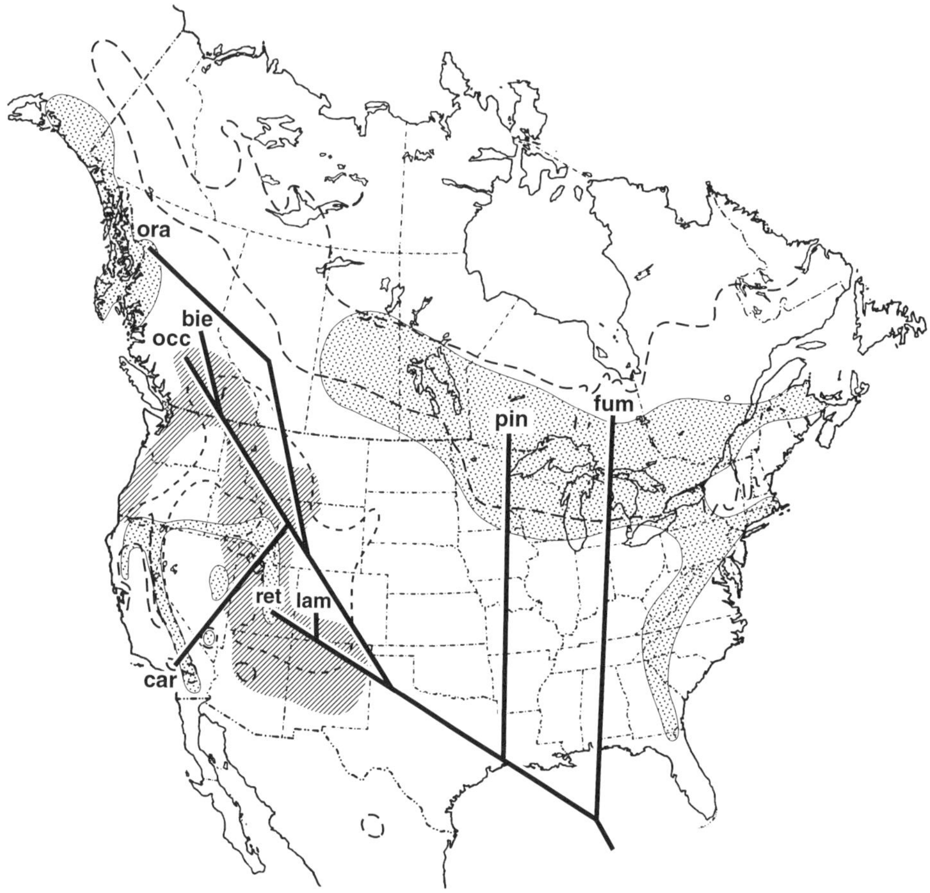

Fig. 7.6. Phylogeography of Nearctic *Choristoneura*: hypothetical cladogram of speciational events (based on Fig. 7.4) superimposed on present day geographical distributions.

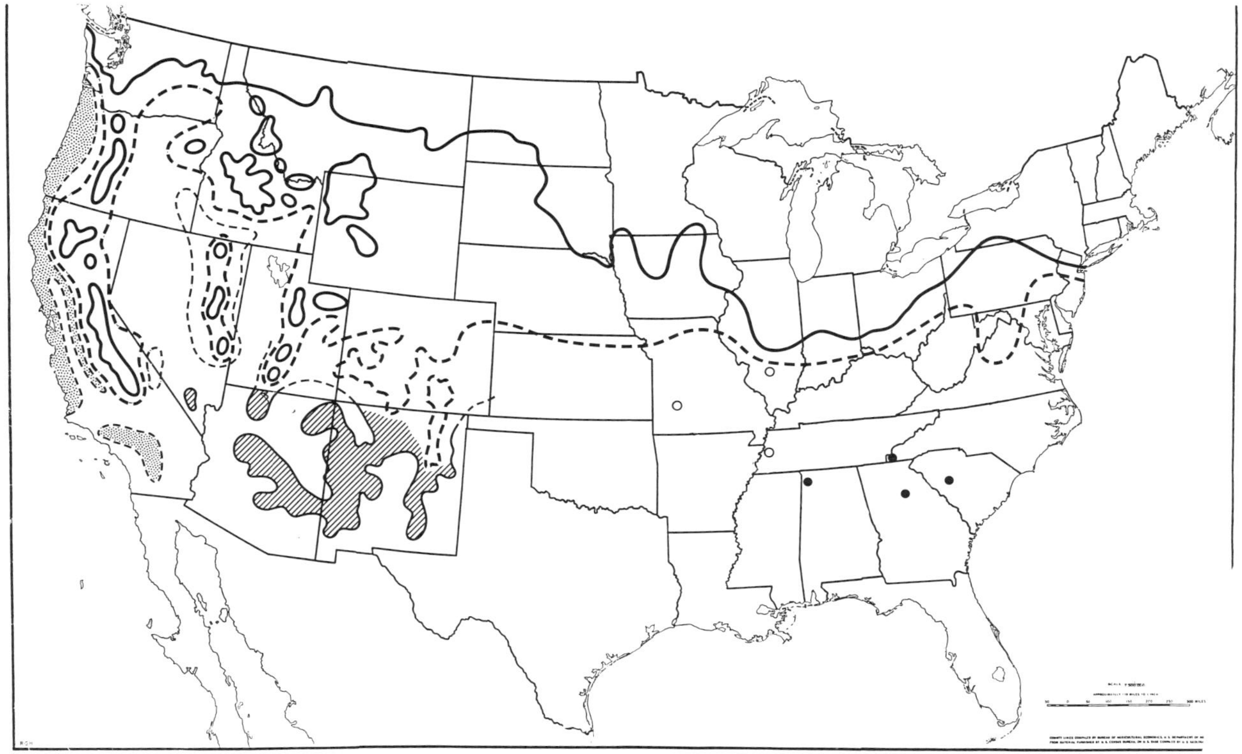

Fig. 7.7. Distribution of coniferous forests in North America during late Pleistocene. Heavy solid lines= approximate extent of Wisconsin glaciation (± 14,000 yr BP); heavy dashed lines= approximate extent of permafrost during maximum glaciation (after Denton and Hughes 1981, Péwé 1983, Porter *et al.* 1983). Stippled area: Pacific Coast Douglas-fir and Big Cone Douglas-fir (17-18,000 yr BP) (after Axelrod and Ting 1961, Axelrod 1966, Baker 1983). Cross-hatched area= Southwestern spruce-fir-pine forest (17-14,000 yr BP) (after Cole 1982, Martin and Mehringer 1965). Open circles= *Picea* (23-21,000 yr BP); closed circles= *Abies balsamea* (20-16,000 yr BP).

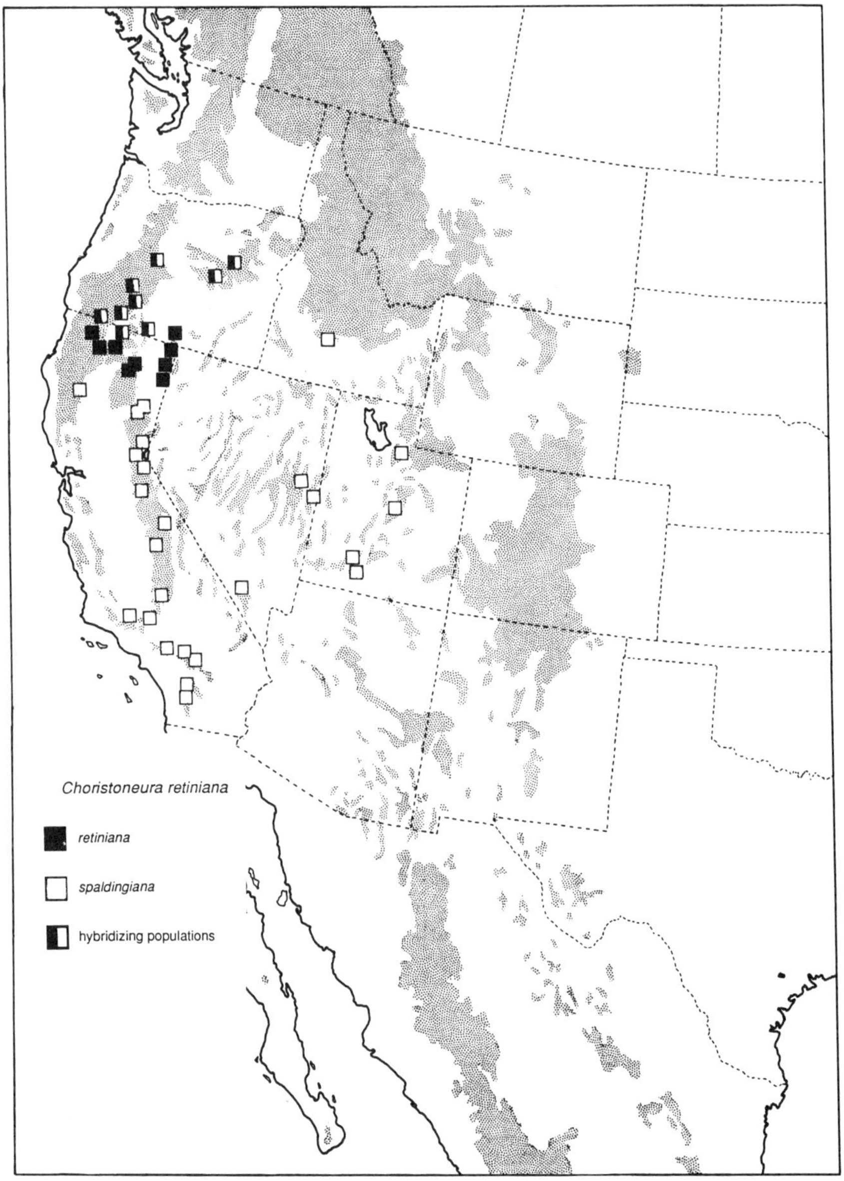

Fig. 7.8. Geographic distribution of *Choristoneura retiniana* based on confirmed collection records. Pheromone attraction in Oregon (half-filled squares) failed to segregate recognizably distinct populations of *C. retiniana* and *C. occidentalis*. Populations assigned to *C. r. retiniana* in northern California (filled squares) probably represent the southern limit of this hybrid zone. Populations of *C. retiniana spaldingiana* (open squares) are much less variable.

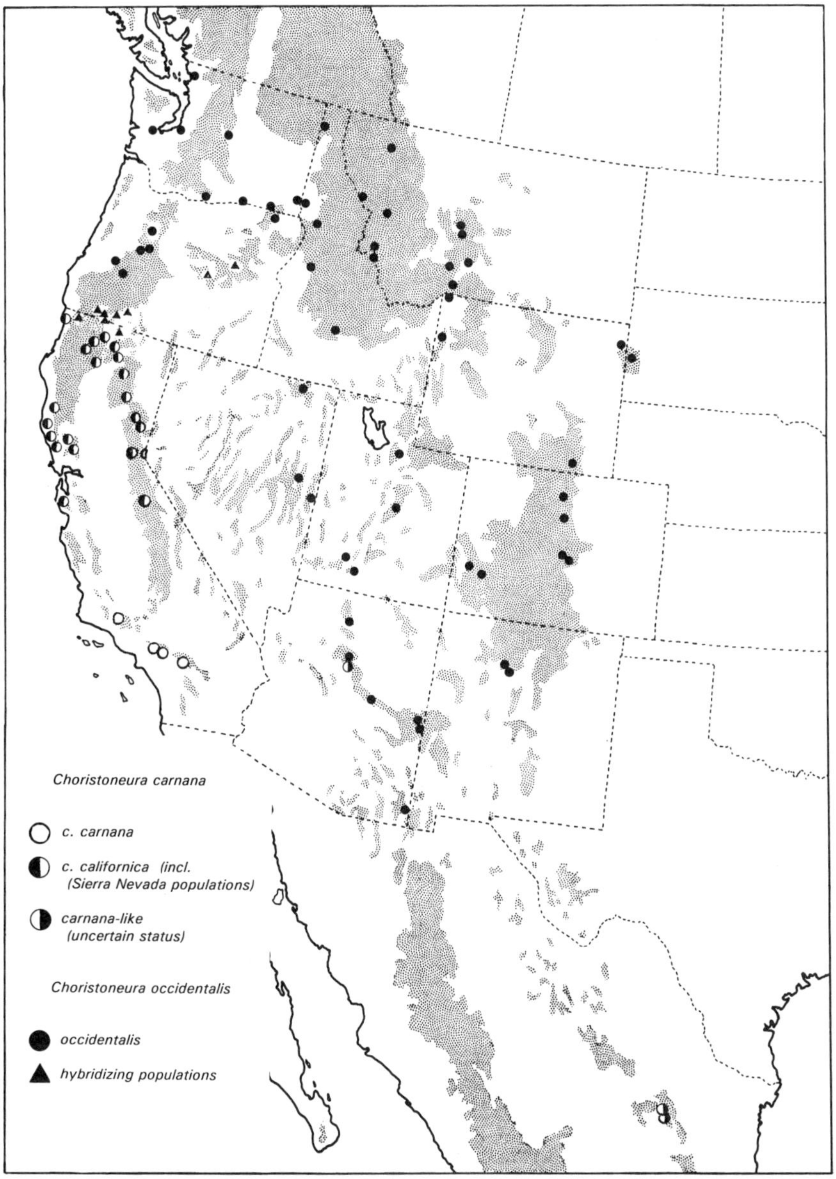

Fig. 7.9. Geographic distribution of *Choristoneura carnana* and *C. occidentalis* (filled circles), based on confirmed collection records. Populations of southern and east-central Oregon (triangles) are hybridizing with *C. retiniana*. Populations of the Sierra Nevada are regarded as atypical *C. carnana californica* (left half-filled circles), intermediate to *C. c. carnana* (open circles). Populations similar to *C. carnana* in northern Arizona and Coahuila, Mexico (right half-filled circles), are of uncertain relationships.

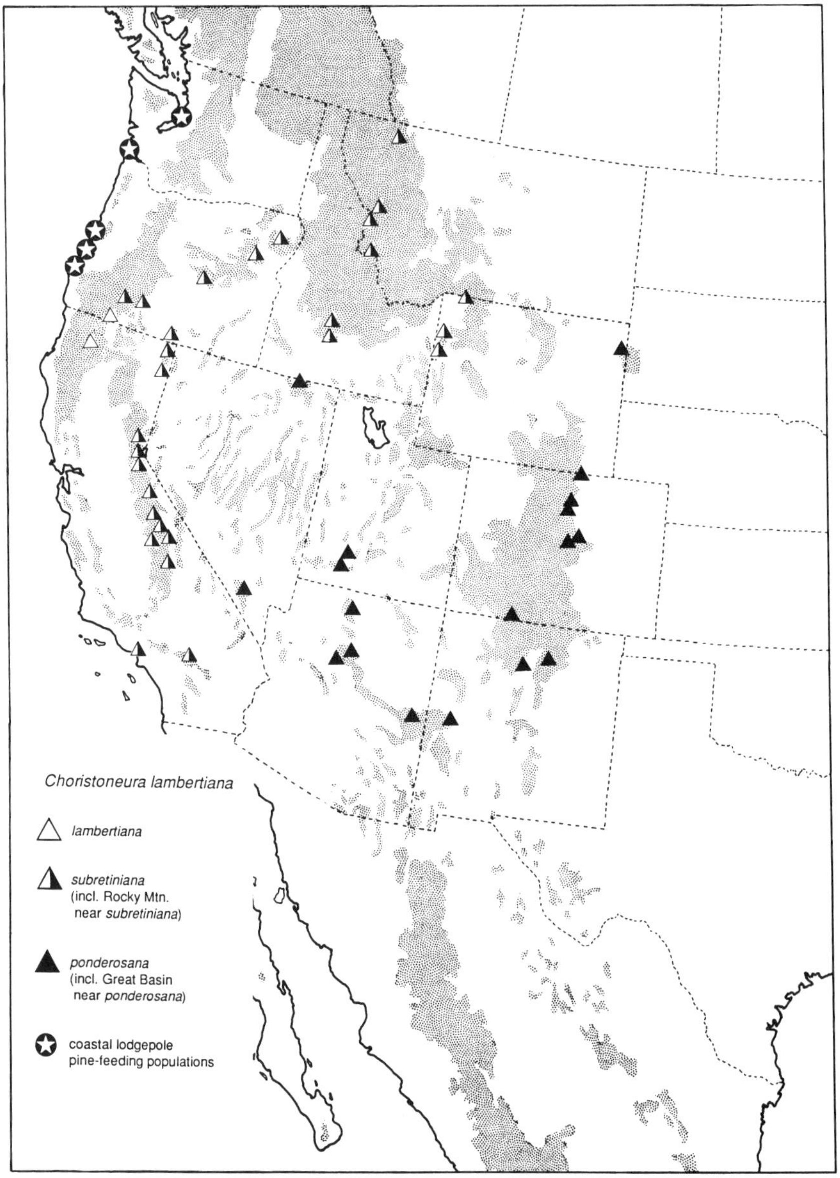

Fig. 7.10. Geographic distribution of *Choristoneura lambertiana*. Populations in central and eastern Oregon, Idaho, western Montana, and Wyoming (half-filled triangles) are intermediate phenotypically between *C. l. lambertiana* (open triangles), typical *C. l. subretiniana* in California, and the Great Basin populations of *C. l. ponderosana* (filled triangles), which in turn are intermediate to the typical phenotype of *C. l. ponderosana* in the eastern Rocky Mountains. The coastal populations in Oregon and Washington (circle stars) may represent a western segregate of *C. pinus*, rather than a component of *C. lambertiana*.

Figs. 7.11-7.82. Voucher specimens for photographs are in UCB; all De Benedictis and Powell collectors, except as noted otherwise.

Figs. 7.11-7.16. *Choristoneura retiniana* (Walsingham), Mt. Shasta, Siskiyou Co., Calif. 7.11. Holotype male, "Mt. Shasta, Siskiyou Co., California, 2.VIII-1.IX-1871" (Walsingham) (BMNH). 7.12. Male, McBride Spr., 5000', VII-21-66, at light. "Comp. w/TYPE BM, JAP '71." 7.13-14. Males, Everitt Memorial, 6000', VI-24-74, r.f. *Abies concolor* (JAP 74F58) (T.D. Eichlin and J. Powell). 7.15-16. Females, same data as 7.13-14.

Figs. 7.17-7.20. *Choristoneura retiniana retiniana* (Walsingham), Salmon Mts., Siskiyou Co., Calif. 7.17-18. Males, Trail Creek, 17 air km SW of Callahan, VII-27-82, blacklight. 7.19. Female, Hockaday Spring, 5 km SW Etna, 4600', VI-21-78, r.f. *Abies* (Volney and Waters). 7.20. Female, same locality, VI-24-80, r.f. *A. concolor* (JAP 80F64).

11
12
13
14
15
16
17
18
19
20

Figs. 7.21-7.28. *Choristoneura retiniana retiniana* (Walsingham), Modoc Co., Calif. (all r.f. *Abies concolor*, 1974 collections by USFS personnel). 7.21. Male, 3 mi S Cedar Pass, VII-6-74 (JAP 74G36). 7.22. Male, 17 mi S Cedar Pass, VII-9-74 (JAP 74G40). 7.23. Male, 6 mi NW Cedar Pass, VI-30-74 (JAP 74G16). 7.24. Male, Rush Cr., 9 mi NE Adin, VI-12-74 (JAP 74F25). 7.25. Female, 12 mi S Cedar Pass, V-24-74 (JAP 74F70). 7.26. Female, 17 mi S Cedar Pass, V-24-74 (JAP 74F72). 7.27. Female, 10 mi N Cedar Pass, VII-7-74 (JAP 74G32). 7.28. Female, 6 mi NW Cedar Pass, VII-7-74 (JAP 74G34).

21
22
23
24
25
26
27
28

Figs. 7.29-7.32. *Choristoneura retiniana spaldingiana* Obraztsov, Tehachapi Mts., Calif. 7.29. Male, Tehachapi Mt. Park, Kern Co., 5,000', VI-16-81, r.f. *Abies concolor* (JAP 81F101.2). 7.30. Male, same data (JAP 81F104). 7.31-32. Females, same locality, VI-16-80, r.f. *A. concolor* (JAP 80F25).

29
30
31
32

Figs. 7.33-7.36. *Choristoneura carnana carnana* (Barnes and Busck), San Gabriel and San Bernardino Mts., Calif. 7.33-35. Males, 4 km NE Mt. Baldy P.O., 6,600', San Bernardino Co., VI-17-80, r.f. *Pseudotsuga macrocarpa* (JAP 80F29). 7.36. Female, 1 mi E Angeles Oaks, San Bernardino Co., VI-24-82, r.f. *P. macrocarpa* (JAP 82F18).

Figs. 7.37-7.40. *Choristoneura carnana* (Barnes and Busck), Sierra Nevada populations. 7.37. Male, Blodgett Forest, 13 mi E Georgetown, 4200', El Dorado Co., VI-24-81, r.f. *Pseudotsuga menziesii* (JAP 81F116). 7.38. Female, same locality, VIII-3-80, r.f. *P. menziesii* (JAP 80F108). 7.39-40. Males, Butt Valley Res., Plumas Co., VII-4-65 (D. and K. Rentz).

Figs. 7.41-7.46. *Choristoneura carnana californica* Powell, typical populations. 7.41. Male paratype, Anderson Spr., Lake Co., VI-24-58 (W. Bauer and S. Buckett). 7.42. Male, R. L. Stevenson Park, Mt. St. Helena, Napa Co., VI-5-80, r.f. *Pseudotsuga menziesii* (JAP 80F7). 7.43. Male, 1 mi NW Cobb, Lake Co., VI-5-80, r.f. *P. menziesii* (JAP 80F12). 7.44-45. Females, 1 mi SE Angwin, Napa Co., V-20-80, r.f. *P. menziesii* (JAP 80E42). 7.46. Female, same data as Fig. 7.42.

Figs. 7.47-7.50. *Choristoneura carnana californica* Powell, Trinity-Salmon Mts., Calif. 7.47. Male, Nelson Cr. Gap, 3600', 6 km ESE Trinity Center, Trinity Co., VI-23-80, r.f. *Pseudotsuga menziesii* (JAP 80F49). 7.48. Male, Blue Mtn., 4600', 10 air km SE Trinity Center, Trinity Co., VI-23-80, r.f. *P. menziesii* (JAP 80F43). 7.49. Female, same data as 7.47. 7.50. Female, same data as 7.48.

41
42
43
44
45
46
47
48
49
50

Figs. 7.51-7.56. *Choristoneura carnana/retiniana* hybrid population, Mt. Ashland, Jackson, Co., Oregon. 7.51-54. Males, 3 air km S Ashland, VII-28-82, blacklight. 7.55. Male, W branch Long John Cr., VI-22-78, r.f. *Abies* (Volney and Waters). 7.56. Female, same data as 7.51-54.

51
52
53
54
55
56

Figs. 7.57-7.62. *Choristoneura retiniana/occidentalis* hybrid population, Aldrich-Strawberry Mts., Grant Co., Ore. 7.57. Male, Starr Campgr., 15 air km SW Canyon City, VIII-2-82, r.f. *Abies grandis* (JAP 82H2) (green pupa). 7.58. Male, same data (JAP 82H6) (intermediate brown larva). 7.59. Male, same data, attracted to "OXY" lure. 7.60-61. Females, same data as 7.57. 7.62. Female, same data (JAP 82H8) (green larva).

57
58
59
60
61
62

Figs. 7.63-7.68. *Choristoneura occidentalis* Freeman, eastern Cascade Range, Oganogan Co., Washington. 7.63. Male, 7 mi NW Conconully, 3500', VII-3-88, r.f. *Pseudotsuga menziesii* (JAP 88G11). 7.64-65. Males, 12 mi NW Conconully, 4500', VII-3-88, r.f. *P. menziesii* (JAP 88G15). 7.66. Female, same data as 7.64-65. 7.67-68. Females, 10 mi NW Conconully, 4000', VII-3-88, r.f. *P. menziesii* (JAP 88G14).

Figs. 7.69-7.70. *Choristoneura occidentalis* Freeman, Schell Creek Range, White Pine Co., Nevada. 7.69. Male, Timber Cr., 7 air mi E McGill, 8800', VIII-5-78, blacklight. 7.70. Female, same locality, VIII-7-80, blacklight.

Figs. 7.71-7.72. *Choristoneura carnana/occidentalis* complex, northern Mexico. 7.71. Male, 6 mi E Los Lirios, 7540', Coahuila, VII-10-86, blacklight (P. Jump). 7.72. Female, Las Alazanas, Coahuila, VI-22-81 [r.f.] *Pseudotsuga macrolepis*, 761-B (D. Cibriari, T. Mendez).

63
64
65
66
67
68
69
70
71
72

Figs. 7.73-7.78. *Choristoneura lambertiana* (Busck). 7.73. *C. lambertiana lambertiana*, male, Sawyer's Bar, Klamath Natl. Forest, Siskiyou Co., Calif., VI-7-58, r.f. *Pinus lambertiana* (C. R. Quick). 7.74. *C. lambertiana* complex, male, Tugman St. Park nr Lakeside, Coos Co., VIII-10-76, at light. 7.75. *C. lambertiana subretiniana* Obraztsov, male, Donner Camp, N of Truckee, Nevada Co., Calif., VII-10-80, r.f. *Pinus contorta* (JAP 80G15). 7.76. *C. l. subretiniana*, female, Sagehen Creek 6800', Nevada Co., Calif., VII-8-70, r.f. *Abies concolor* (JAP 70G22). 7.77-78. *C. lambertiana* complex, western Montana: 7.77, Male, Bonner, Missoula Co., VI-30-69, r.f. *Pinus contorta* (McGregor and Shea); 7.78, female same data.

Figs. 7.79-7.82. *Choristoneura lambertiana ponderosana* Obraztsov. 7.79. Male, 5 km W Pagosa Springs, Archuleta Co., Colorado, VII-18-73. [r.f.] *Pinus ponderosa* (Hopk. 36745, R. Stevens). 7.80. Female, Sugarloaf [Boulder Co.], Colo., "7-8-37" [r.f.] *Pinus ponderosa* (Hopk. 17701-B, J.A. Beal). 7.81. Male, Los Alamos, Santa Fe Co., New Mexico, VI-20-81, r.f. *P. ponderosa* (A. Liebhold). 7.82. Male, Duck Cr., 5 km E Navajo Lake, Kane Co., Utah, VII-23-81, blacklight.

73
74
75
76
77
78
79
80
81
82